DIGITAL LOGIC CIRCUITS: TESTS AND ANALYSIS

Bob Middleton has been a professional free-lance technical writer in the electronic field for many years. His many books and magazine articles have proved invaluable to students, technicians, and engineers, because they have all been based on his own practical experience. His home workshop is filled with test equipment, receivers, and other equipment that he uses to check out every detail in preparing his many books.

Other Sams books by Mr. Middleton include *101 Ways to Use Your VOM, TVM & DVM, Know Your Oscilloscope, Troubleshooting With the Oscilloscope, Effectively Using the Oscilloscope,* and many others.

DIGITAL LOGIC CIRCUITS: TESTS AND ANALYSIS

BY
ROBERT G. MIDDLETON

Howard W. Sams & Co., Inc.
4300 WEST 62ND ST. INDIANAPOLIS, INDIANA 46268 USA

International Standard Book Number: 0-672-21799-6
Library of Congress Catalog Card Number: 81-86555

Edited by: *Welborn Associates*
Illustrated by: *Christine Lixon*

Printed in the United States of America.

PREFACE

Troubleshooting digital logic circuits is now an activity of major concern to electronic technicians. A *digital revolution* has occurred and has irreversibly changed the structure of consumer electronics. For example, television receiver tuning is often accomplished by digital arrangements, and specialized models provide for extensive digital pre-programming of receiver operation. Digital controlled readout of the operating channel and the time may also be provided. Digital filtering techniques are utilized for optimum separation of chroma and Y signals. Scanner-monitor radio receivers employ automatic digital tuning techniques. A frequency readout is provided on am, fm, and communications receivers. Digital control techniques and programming routines have made significant entry in the high-fidelity scene. Personal computers have found widespread acceptance in homes and offices. Automobiles are becoming computerized to an ever-increasing extent.

This book emphasizes practice—not theory. Its purpose is to show you how to make basic digital tests and measurements as efficiently as possible. The techniques are presented without frills or double talk. Most digital servicing is done with logic probes and logic pulsers. These instruments are in the category of go/no-go testers, although they also provide general test data concerning frequency, transient occurrences, and logic levels. Logic probes and pulsers are voltage-operated devices. They are supplemented advantageously by the current tracer which senses current flow in the circuit. Logic clips, comparators, and oscilloscopes are also used in practical test procedures. Sometimes malfunction in a digital system results merely from a marginal power-supply voltage—here, the dvm does yeoman service. Even the ohmmeter finds occasional application.

In this book, no previous experience with digital circuitry is assumed. The coverage starts with simple gates, illustrates commercial IC package pinouts, and introduces the reader to truth tables. Test procedures are initially concerned with verifying device input/output relations as specified in the associated truth table. Of course, not all digital trouble symptoms are caused by IC faults—poor contacts, solder "whiskers," or broken pc conductors may also cause trouble symptoms. Suitable follow-up tests are often required to pinpoint the trouble to device failure or to external circuitry. We have attempted to anticipate all of the pitfalls that lie in wait for the beginner. It is sometimes remarked that a digital troubleshooter need pay attention only to the IC package pinout and to its truth table. However, this is somewhat of an oversimplification. As in television troubleshooting procedures, the technician will be called upon to reason back from symptom to cause, and to systematically eliminate various possibilities on the basis of tests and measurements.

Electronics technology is becoming increasingly advanced and sophisticated. We must keep up with these advances if we are to remain competitive. Unless the full capabilities of the logic probe, pulser, current tracer, logic clip, comparator, and oscilloscope are clearly understood, it will become much more difficult in the future to properly service modern electronic circuitry.

In preparing this new working handbook, I have recognized the current need, and have made a dedicated effort to meet it. It is my firm belief and sincerest hope that this book will be a valuable addition to the libraries of all present and future electronic service technicians.

ROBERT G. MIDDLETON

CONTENTS

PART 1

INTRODUCTION

Section 1

Logic Pulser—Basic Gates—Common Troubles in Digital Circuitry—Scanner-Monitor Digital-Logic Section Checks—Positive Logic and Negative Logic—Applications of Logic Current Tracer—Equivalent Gates in Digital Logic Circuits—Common Causes for Puzzling Observations—Sensitivity Control Settings—Digital IC Failure Modes—Waveform Checks in Dedicated Digital Equipment—Digital Pulse Voltage Tolerance—Level Indicators—Current and Voltage in Failed ICs—In-Circuit Versus Out-of-Circuit Tests—Fan-In/Fan-Out Principles—Glitches Caused by Races at Gate Inputs—Noise Glitches in Digital Waveforms

Section 2

Half-Adder Operation—Operation of Basic Full Adder—Subtraction by Full Adders—Parallel Adder Function—Sign-Magnitude Adder/ Subtracter

Section 3

Section 4

Section 5

Section 6

Section 7

Section 8

Section 9

Section 10

Section 11

Section 12

Section 13

Section 14

PART 2

DIGITAL TROUBLESHOOTING

TESTS

Section 15

Section 16

Section 17

Section 18

Section 22

Section 23

Section 24

Section 25

Section 26

Section 27

Section 28

Part 1

INTRODUCTION

Section 1
BASIC TEST EQUIPMENT

The most basic and most important unit of digital and logic test equipment is the digital logic probe. A high-performance logic probe is illustrated in Fig. 1-1. A digital logic probe is an indicator of logic activity (or lack of activity). For example, the Hewlett-Packard HP-545A logic probe contains a probe-tip indicator lamp with four types of indication.

Fig. 1-1. Logic probe in use. *(Courtesy Hewlett-Packard Co.)*

1. Lamp glows at full brightness—probe tip is applied to a logic-high point (see Fig. 1-2).
2. Lamp is dark—probe tip is applied to a logic-low point.
3. Lamp glows at half-brightness—probe tip is applied to an open circuit, or "bad level" (see Fig. 1-3).
4. Lamp flashes—probe tip is applied to a pulsing point; a single pulse produces one flash, and a pulse train produces 10 flashes per second.

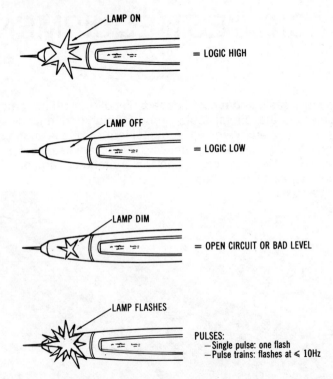

LAMP ON

= LOGIC HIGH

LAMP OFF

= LOGIC LOW

LAMP DIM

= OPEN CIRCUIT OR BAD LEVEL

LAMP FLASHES

PULSES:
—Single pulse: one flash
—Pulse trains: flashes at ≤ 10Hz

Fig. 1-2. Logic probe indicator action. *(Courtesy Hewlett-Packard Co.)*

This type of probe also contains a pulse memory, whereby a single pulse can be "caught" and "stretched" for reliable indication, although the pulse width might be only 10 ns (see Fig. 1-4). In addition, this type of probe is provided with a TTL/CMOS switch, whereby tests can be made in either type of circuitry. As shown in Fig. 1-3, a typical TTL logic-1 level is 2.4 volts, and logic-0 level is 0.4 volt, with bad level in between. On the other hand, a typical CMOS logic-1 level is 4.2 volts, and logic-0 level

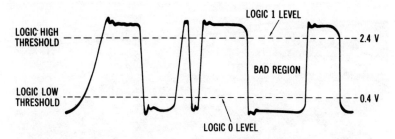

Fig. 1-3. Logic high, logic low, and bad region for TTL circuitry.

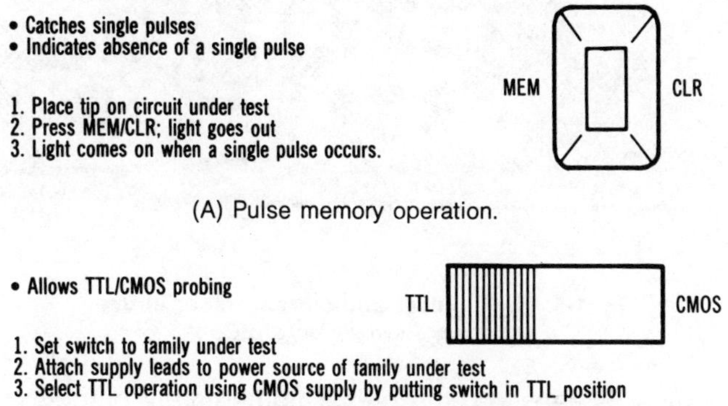

- Catches single pulses
- Indicates absence of a single pulse

1. Place tip on circuit under test
2. Press MEM/CLR; light goes out
3. Light comes on when a single pulse occurs.

MEM   CLR

(A) Pulse memory operation.

- Allows TTL/CMOS probing

TTL   CMOS

1. Set switch to family under test
2. Attach supply leads to power source of family under test
3. Select TTL operation using CMOS supply by putting switch in TTL position

(B) TTL/CMOS switch operation.

Fig. 1-4. Logic probe contains pulse memory, with switch for TTL or CMOS operation. *(Courtesy Hewlett-Packard Co.)*

is 1.8 volts, with bad level in between. When tests are made in-circuit, each IC terminal will be either *high* or *low,* unless a fault causes a bad level condition. When tests are made out-of-circuit, all IC input pins that are not being driven should be returned to ground (or to a *high* potential, if test conditions so dictate).

Logic Pulser

The second most basic and important unit of digital and logic test equipment is the digital logic pulser. A high-performance logic pulser is illustrated in Fig. 1-5.

Pulser Output Waveform

Note that when the push button of the Hewlett-Packard HP-546A

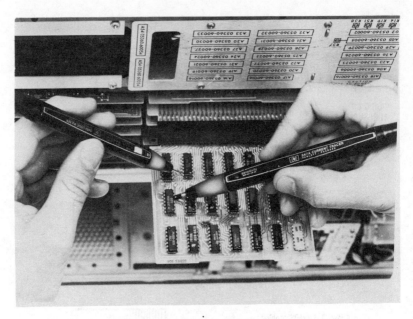

Fig. 1-5. Logic pulser and current tracer in use.
(Courtesy Hewlett-Packard Co.)

logic pulser is pressed, the pulser immediately outputs a single dual-polarity pulse, as depicted in Fig. 1-6. This is the waveform that is produced when pulsing into an open circuit. First, the waveform goes *low,* then pauses and goes *high.*

Automatic Pulse Width, Height, and Polarity

The foregoing diagram shows the maximum pulse width provided by this type of pulser. Note that when the pulse is injected into a logic circuit, the current flow through the tip of the pulser is sensed by an output sensing circuit that turns the pulser off. The total test energy is kept small, to eliminate any possibility of damage to the circuit being pulsed. The pulse height or amplitude is determined by the power-supply voltage that is used. Therefore, the pulser should be powered from the circuit under test, or from a power supply that provides the same voltage.

Pulsing Into a Load

When this type of pulser drives a *low* node (interconnection between ICs) to a *high* state, the pulser produces no change until it drives the circuit *high.* When the pulser goes *high,* it supplies

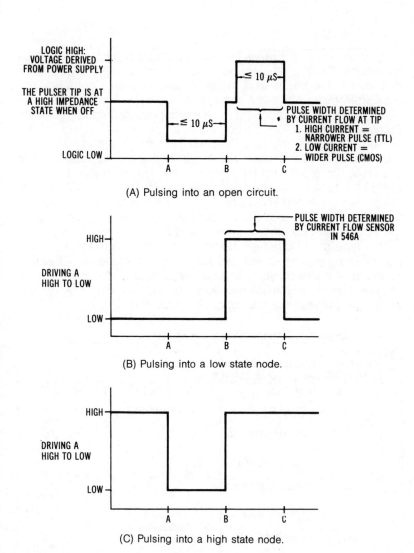

(A) Pulsing into an open circuit.

LOGIC HIGH: VOLTAGE DERIVED FROM POWER SUPPLY

THE PULSER TIP IS AT A HIGH IMPEDANCE STATE WHEN OFF

LOGIC LOW

≤ 10 μS

≤ 10 μS

PULSE WIDTH DETERMINED BY CURRENT FLOW AT TIP
1. HIGH CURRENT = NARROWER PULSE (TTL)
2. LOW CURRENT = WIDER PULSE (CMOS)

A   B   C

(B) Pulsing into a low state node.

HIGH

DRIVING A HIGH TO LOW

LOW

PULSE WIDTH DETERMINED BY CURRENT FLOW SENSOR IN 546A

A   B   C

(C) Pulsing into a high state node.

HIGH

DRIVING A HIGH TO LOW

LOW

A   B   C

Fig. 1-6. Logic pulser waveforms. *(Courtesy Hewlett-Packard Co.)*

sufficient output drive to take any normal digital circuit or bus *high* momentarily. On the other hand, when this type of pulser drives a *high* node to a *low* state, the injected waveform pulses the *high* state to a *low* state for the duration of the test pulse. The node then returns to its initial *high* state, and the *high* portion of the pulser output has no effect on the circuit. Note that a node is an interconnection, such as a pc conductor, between IC package terminals.

23

Output Modes

A basic logic pulser is the equivalent of a single-shot pulse generator. On the other hand, an elaborated logic pulser also provides pulse bursts and pulse streams. As an illustration, the Hewlett-Packard HP 546A logic pulser provides six push-button programmable output modes, automatic polarity output, automatic pulse width, and automatic pulse amplitude. In turn, a gate can be pulsed in single steps, a logic circuit can be driven at 100 Hz, or a device such as a counter can be loaded with a precise number of pulses. Programming facilities for this type of pulser are listed in Chart 1-1. Note that the first pulse is outputted after subtracting the pulses produced when programming the output. Note also that the pulse button should be released *during* the final burst. In turn, the pulser will complete the burst and then shut off.

This type of logic pulser includes an LED to indicate when the pulser is operating. It verifies the program mode and permits counting of pulse bursts. For example, 256 output pulses can be obtained by programming two bursts of 100, five bursts of 10, and six single pulses (see tabulation in Chart 1-1).

Chart 1-1. Logic Pulser Output Programming Modes

OPERATION:			
PRESS AND RELEASE CODE BUTTON o			
PRESS AND LATCH CODE BUTTON o_			
OUTPUT MODES:		**TO OUTPUT EXACTLY 432 PULSES:**	
o	SINGLE PULSE	1. 100 Hz BURST oo_	98^1
			100
o_	100 Hz STREAM		100
			100^2
oo_	100 Hz BURST		400
		2. 10 Hz BURST oooo_	6^1
ooo_	10 Hz STREAM		10
			10^2
oooo_	10 Hz BURST		430
		3. SINGLE PULSE o	1
ooooo_	1 Hz STREAM	SINGLE PULSE o	1
			432

Courtesy Hewlett-Packard Co.

Basic Gates

At this point, it is helpful to briefly review basic gate functions. As shown in Fig. 1-7, an AND gate produces an output only if all of its inputs are *high.* An OR gate (Fig. 1-8) produces an output if any

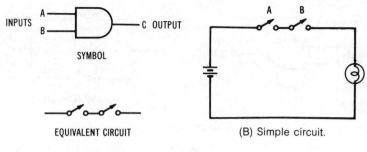

(A) Logic symbol.

(B) Simple circuit.

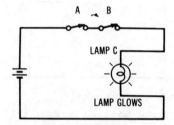

NOTE: If either switch A or switch B (or both) is opened, the lamp will be dark.

(C) Both switches closed.

C = A AND B
C = A · B = AB

INPUTS		OUTPUT
A	B	C
0	0	0
0	1	0
1	0	0
1	1	1

(D) Truth table.

1 = HIGH = H = TRUE

0 = LOW = L = FALSE

(E) Alternative logic-state notations.

Fig. 1-7. A 2-input AND gate.

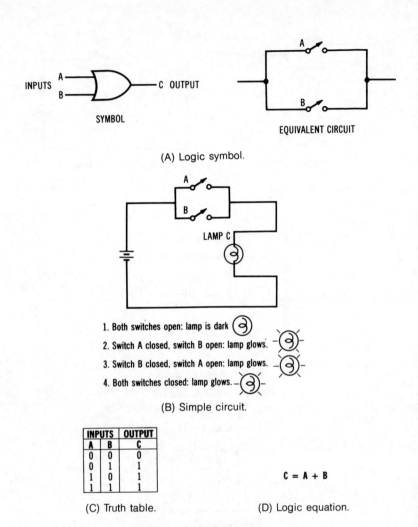

(A) Logic symbol.

(B) Simple circuit.

1. Both switches open: lamp is dark
2. Switch A closed, switch B open: lamp glows.
3. Switch B closed, switch A open: lamp glows.
4. Both switches closed: lamp glows.

INPUTS		OUTPUT
A	B	C
0	0	0
0	1	1
1	0	1
1	1	1

(C) Truth table.

$$C = A + B$$

(D) Logic equation.

Fig. 1-8. A 2-input OR gate.

one of its inputs is *high*. An XOR gate (Fig. 1-9) produces an output if its inputs are at opposite logic levels; it has no response if its inputs are at the same logic level. Note that a buffer (an amplifier) is not a gate. Similarly, an inverter (an inverting amplifier) is not a gate—an inverter is called an operator. Gates with inverted outputs are shown in Fig. 1-10 through Fig. 1-12.

(A) Logic symbol.

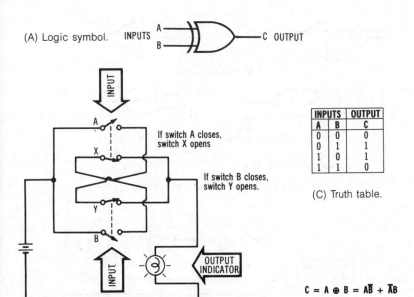

INPUTS A ─⊐⊐)─ C OUTPUT
       B ─⊐⊐)

If switch A closes, switch X opens

If switch B closes, switch Y opens.

INPUTS		OUTPUT
A	B	C
0	0	0
0	1	1
1	0	1
1	1	0

(C) Truth table.

$$C = A \oplus B = A\overline{B} + \overline{A}B$$

(D) Logic equation.

1. If neither switch A nor B is pressed, lamp is dark.
2. If only switch A is pressed, lamp glows.
3. If only switch B is pressed, lamp glows.
4. If both switch A and switch B are pressed, lamp is dark.

(B) Basic equivalent circuit.

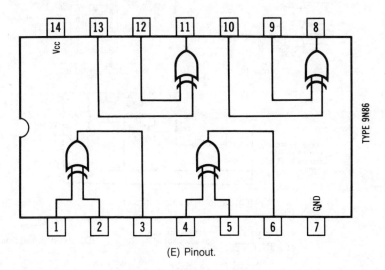

(E) Pinout.

Fig. 1-9. A 2-input XOR gate.

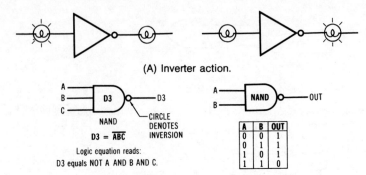

(A) Inverter action.

D3 = $\overline{ABC}$

Logic equation reads:
D3 equals NOT A AND B AND C.

A	B	OUT
0	0	1
0	1	1
1	0	1
1	1	0

(B) Logic symbols and truth table.

Electromechanical NAND action. Relay provides INVERTER function.

(C) Basic equivalent circuit.

Fig. 1–10. Basic gates with inverted outputs.

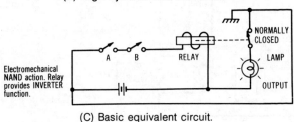

(A) Logic symbols.

A	B	C
1	0	0
0	1	0
1	1	0
0	0	1

(B) Truth table.

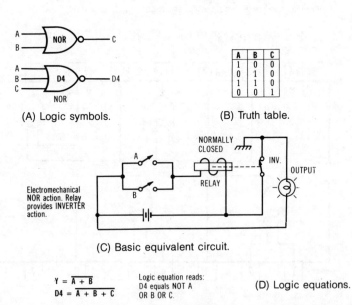

Electromechanical NOR action. Relay provides INVERTER action.

(C) Basic equivalent circuit.

$Y = \overline{A + B}$
$D4 = \overline{A + B + C}$

Logic equation reads:
D4 equals NOT A
OR B OR C.

(D) Logic equations.

Fig. 1–11. NOR gates.

A	B	C
1	0	1
0	1	1
1	1	0
0	0	0

$$C = \overline{A \oplus B}$$
(C = A EXCLUSIVE OR B)

(A) Logic symbol and equation.　　　　(B) Truth table.

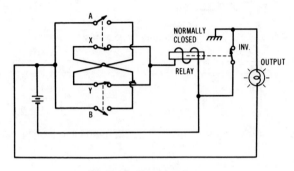

(C) Equivalent circuit.

Fig. 1-12. XOR gate.

Gates With Negated Inputs

As shown in Fig. 1-13, distinction must be made between a NAND gate and a NEGATED AND gate. In other words, a NAND gate consists of an AND gate followed by an inverter, whereas a NEGATED AND gate consists of a pair of inverters followed by an AND gate. Observe in particular that the truth tables are not the same for these two types of gates. The logic equation for the NAND gate is written $Y = \overline{AB}$, whereas the logic equation for the NEGATED AND gate is written $Y = \overline{A} \cdot \overline{B}$, or $Y = \overline{A}\overline{B}$. In turn, the logic equation for the NAND gate is read Y equals not A and B, whereas the logic equation for the NEGATED AND gate is read Y equals not A and not B. Both types of gates are often encountered in logic diagrams.

A NEGATED AND gate may have all of its inputs negated, or it may have only one of its inputs negated. A NEGATED AND gate with four inputs may have two or three of its inputs negated.

An electromechanical equivalent circuit for a 2-input AND gate with one negated input is shown in Fig. 1-14. A comparison of NOR gates and negated OR gates with truth tables is shown in Fig.

29

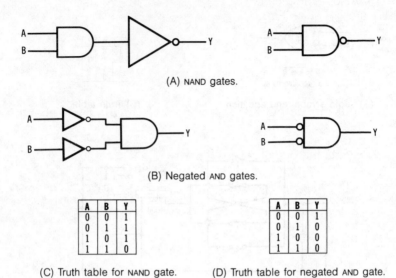

(A) NAND gates.

(B) Negated AND gates.

A	B	Y
0	0	1
0	1	1
1	0	1
1	1	0

A	B	Y
0	0	1
0	1	0
1	0	0
1	1	0

(C) Truth table for NAND gate.   (D) Truth table for negated AND gate.

$$Y = \overline{AB}$$ $$Y = \overline{A}\,\overline{B}$$

(E) Logic equation for NAND gate. . (F) Logic equation for negated AND gate.

Fig. 1-13. Comparison of NAND gates and negated AND gates.

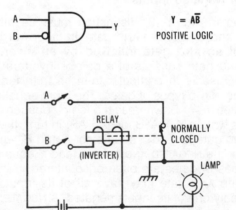

$$Y = A\overline{B}$$
POSITIVE LOGIC

Fig. 1-14. Electromechanical equivalent circuit for a 2-input AND gate with one negated input.

1-15. Also, examples of OR gates with negated AND implementation are shown in Fig. 1-16.

XOR Gate Characteristics

An EXCLUSIVE OR (XOR) gate has two inputs—it produces a logic-high output only when both inputs are simultaneously driven to opposite levels. Thus, if its inputs are 1,1 or 0,0, an XOR gate produces no output (is at a logic-low level). Note that an EXCLUSIVE NOR ($\overline{XOR}$) gate is checked in the same manner as an XOR gate (see Fig. 1-17). The only distinction here is that an $\overline{XOR}$ gate inverts the output that would be obtained with an XOR gate. That is, an $\overline{XOR}$ gate produces a logic-low output only when its inputs are driven to opposite levels (see Fig. 1-18). Thus, if the XOR inputs are 1,1 or 0,0, that gate output is logic high. Normal responses of AND, OR, NAND, NOR, XOR, and $\overline{XOR}$ (XNOR) gates are summarized in Fig. 1-19.

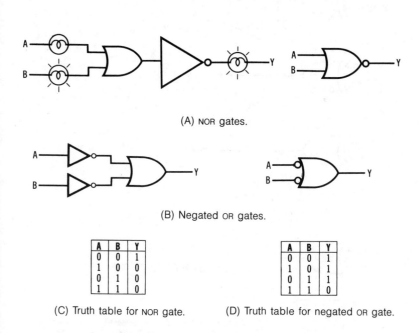

(A) NOR gates.

(B) Negated OR gates.

A	B	Y
0	0	1
1	0	0
0	1	0
1	1	0

(C) Truth table for NOR gate.

A	B	Y
0	0	1
1	0	1
0	1	1
1	1	0

(D) Truth table for negated OR gate.

$$Y = \overline{A + B}$$

(E) Logic equation for NOR gate.

$$Y = \overline{A} + \overline{B}$$

(F) Logic equation for negated OR gate.

Fig. 1-15. Comparison of NOR gates and negated OR gates.

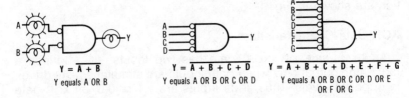

$$Y = \overline{A + B}$$
Y equals A OR B

$$Y = \overline{A + B + C + D}$$
Y equals A OR B OR C OR D

$$Y = \overline{A + B + C + D + E + F + G}$$
Y equals A OR B OR C OR D OR E OR F OR G

Fig. 1-16. Examples of OR gates with negated AND implementation.

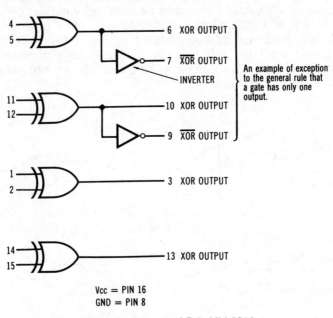

4
5 ──── 6 XOR OUTPUT
7 $\overline{XOR}$ OUTPUT
INVERTER

11
12 ──── 10 XOR OUTPUT
9 $\overline{XOR}$ OUTPUT

An example of exception to the general rule that a gate has only one output.

1
2 ──── 3 XOR OUTPUT

14
15 ──── 13 XOR OUTPUT

Vcc = PIN 16
GND = PIN 8

(A) Logic diagram of Fairchild 9014.

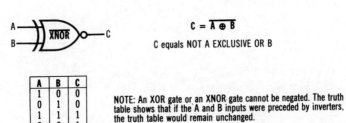

A ──┤ XNOR ├o── C
B ──┤

$$C = \overline{A \oplus B}$$

C equals NOT A EXCLUSIVE OR B

A	B	C
1	0	0
0	1	0
1	1	1
0	0	1

NOTE: An XOR gate or an XNOR gate cannot be negated. The truth table shows that if the A and B inputs were preceded by inverters, the truth table would remain unchanged.

(B) Logic diagram, truth table, and logic equation of XNOR.

Fig. 1-17. XOR/$\overline{XOR}$ and XNOR.

32

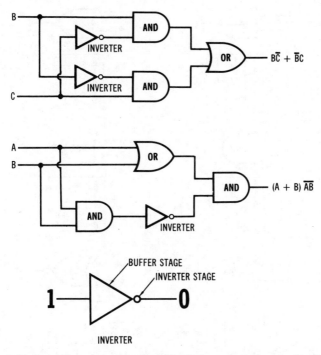

Fig. 1-18. Examples of combinational logic to perform the xor function.

Equivalent XOR Circuitry

Equivalent XOR circuitry may be encountered, and it is helpful for the troubleshooter to recognize these equivalents. One equivalent using AND and OR gates with inverters was shown in Fig. 1-18. Another equivalent is exemplified in Fig. 1-20. The reader can visualize still other equivalents.

AND Gates With One Input

AND gates are manufactured with two or more inputs. However, the digital troubleshooter may encounter logic diagrams that show an AND gate with its inputs short-circuited together, or with only one input, as depicted in Fig. 1-21. In this situation, the AND gate operates as a buffer; a buffer has only one input. Digital circuit designers sometimes use an AND gate for a buffer in this manner, because the gate happens to be "left over" in an IC package. In turn, production costs are lowered by utilizing the AND gate as a buffer.

33

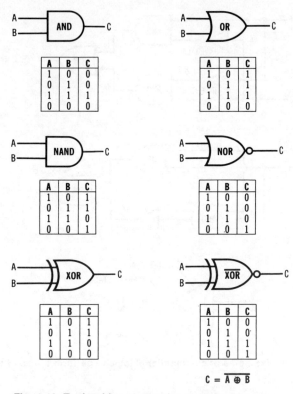

$$C = \overline{A \oplus B}$$

Fig. 1-19. Truth tables summarize normal responses.

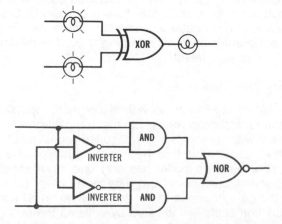

Fig. 1-20. Equivalent logic arrangement for XOR gate function.

Fig. 1-21. AND gates with only one input.

AND-OR Gate Combinations

Various AND-OR gate combinations will be encountered in logic diagrams (Fig. 1-22). The OR gate in Y = AB + CD functions as a "mixer." In other words, if U1 and U2 were tied together, there would be interaction between the gates, and the circuit would not operate properly. However, U3 serves to isolate the outputs from U1 and U2.

AND-OR Gate Equivalents

Previous mention has been made of the AND-OR gate configuration used in logic circuitry. The basic AND-OR gate circuit has two equivalents called the NAND implementation and the NOR implementation, as shown in Fig. 1-23. Other variants can be visualized by the reader in accordance with the principles outlined in Fig. 1-18.

AND-XOR Gate Operation in Basic Adder

As shown in Fig. 1-24 an AND gate is connected with its inputs in parallel with an XOR gate to form a basic adder, called the *half adder*. A half adder has two input terminals, designated A and B. It has two outputs designated S, the *sum* output, and C, the *carry* output. The truth table for the half adder states that the S output will be *high,* provided the inputs A and B are at opposite logic levels, but that the S output will be *low* when the inputs A and B are at the same logic level. The truth table also states that the C output will be *high* provided the inputs A and B are both *high.* In turn, a half adder can add any two binary digits (bits) simultaneously (Table 1-1). Since a half adder has only two input terminals, it cannot add three binary digits simultaneously; this requires three more gates.

XNOR Gate Equivalents

The troubleshooter may encounter equivalent XNOR circuitry occasionally, and it is helpful to recognize these equivalents. One equivalent uses NAND and OR gates, and another equivalent uses AND and OR gates, as shown in Fig. 1-25. The reader will recog-

35

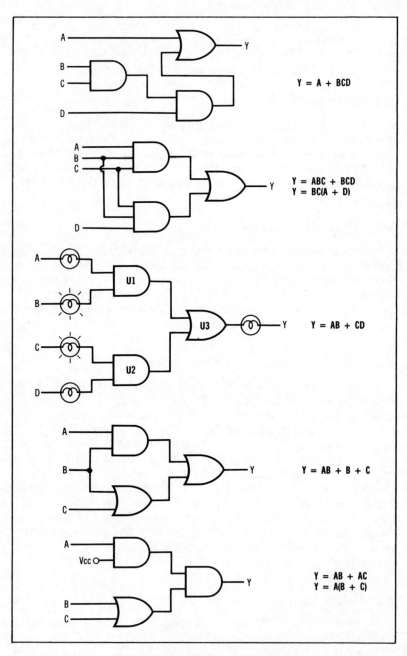

Fig. 1-22. AND-OR gate combinations.

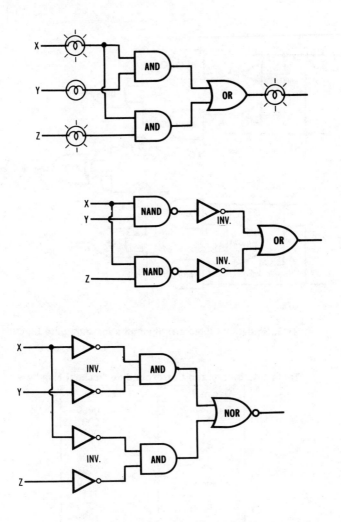

Fig. 1-23. Equivalent AND-OR gate circuitry.

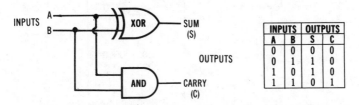

INPUTS		OUTPUTS	
A	B	S	C
0	0	0	0
0	1	1	0
1	0	1	0
1	1	0	1

Fig. 1-24. Connection of AND to XOR gate to form a half adder.

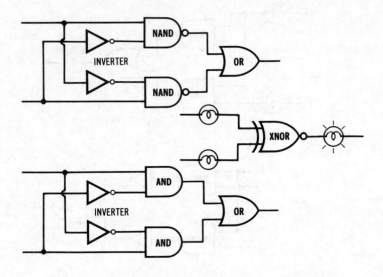

Fig. 1-25. Equivalent logic arrangements for XNOR gate function.

Table 1-1. Binary and Corresponding Decimal Numbers

Decimal	Binary
0	0000
1	0001
2	0010
3	0011
4	0100
5	0101
6	0110
7	0111
8	1000
9	1001
10	1010
11	1011
12	1100
13	1101
14	1110
15	1111
16	0001 0000
17	0001 0001
30	0001 1110
31	0001 1111

nize that there are still other equivalents, in accordance with the principles outlined in Fig. 1-18.

Unused Gate Inputs

It was previously noted that unused gate inputs should not be permitted to "float." This precaution is taken to minimize noise sensitivity and to optimize switching time. Thus, unused inputs of all TTL circuits should be held between 2.4 volts and the absolute maximum of 5.5 volts. Possible ways of handling unused inputs are to connect the unused inputs to a used input, if maximum *high* level fan-out of the driving output will not be exceeded. Each additional input presents a full load to the driving output at a *high* level voltage, but adds no loading at a *low* voltage level.

The high-level fan-out for all circuits is customarily specified at double the low-level fan-out particularly to provide for this method of treating unused inputs. Alternatively, the unused inputs can be connected to V_{cc} through a 1 kilohm resistor. If a transient exceeding the 5.5-volt maximum rating should occur, the impedance will be sufficiently high to protect the input. From 1 to 25 unused inputs may be connected to each 1-kilohm resistor. Again, the unused inputs may be tied to the output of an unused gate in the system. The gate must provide a constant high level output. Still another method is to connect unused inputs to an independent supply voltage in the range of 2.4 to 3.5 volts.

Ground and V_{cc} Terminals

Every gate has a V_{cc} and a ground terminal, as shown in Fig. 1-26. If the power supply and ground lines are not indicated in a logic diagram, they are implied. The digital troubleshooter must be alert to the possibility of internal shorts in an IC package. An internal short might be to ground, or it might be to V_{cc}. Sometimes V_{cc} shorts to ground.

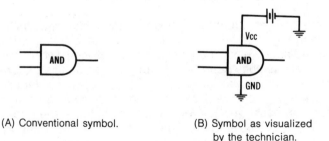

(A) Conventional symbol.

(B) Symbol as visualized by the technician.

Fig. 1-26. AND gate symbols.

39

Summary of Basic Gate Operation

At this point, it is helpful to summarize the responses of AND, OR, NAND, and NOR gates, as depicted in Fig. 1-27. It is seen that the NOR gate is an "upside down" AND gate; similarly, a NAND gate is an "upside down" OR gate. An AND gate is the "reverse" of an OR gate (an AND gate "looks like" an OR gate when operating in a negative-logic system). Again, a NAND gate is the "reverse" of a NOR gate; (a NAND gate "looks like" a NOR gate when operating in a negative-logic system).

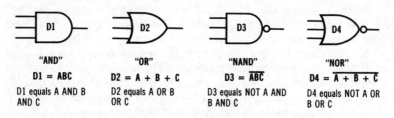

	"AND"	"OR"	"NAND"	"NOR"
	D1 = ABC	D2 = A + B + C	D3 = $\overline{ABC}$	D4 = $\overline{A + B + C}$
	D1 equals A AND B AND C	D2 equals A OR B OR C	D3 equals NOT A AND B AND C	D4 equals NOT A OR B OR C

INPUT STATES			CORRESPONDING OUTPUT STATE			
A	B	C	D1	D2	D3	D4
0	0	0	0	0	1	1
1	0	0	0	1	1	0
0	1	0	0	1	1	0
1	1	0	0	1	1	0
0	0	1	0	1	1	0
1	0	1	0	1	1	0
0	1	1	0	1	1	0
1	1	1	1	1	0	0

Fig. 1-27. Basic gate symbols, logic equations, and truth table for AND, OR, NAND, and NOR gates. *(Courtesy Hewlett-Packard Co.)*

Common Troubles in Digital Circuitry

Some common troubles that occur in digital circuitry are depicted in Fig. 1-28. For example, if there is a solder bridge between adjacent pins, there is a short between two conductors. Conversely, a faulty bond (joint) can produce an internal open circuit condition. Or, an internal short may occur, as depicted between the V_{cc} lead and an adjacent AND gate input. A solder bridge can cause a *stuck-low* trouble symptom—the node is *low* and cannot be driven *high*. Again, an internal short can cause a *stuck-high* trouble symptom—the node is *high* and cannot be driven *low*. A logic probe permits the technician to "see" stuck

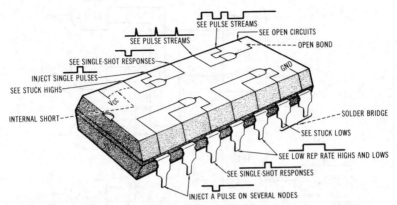

Fig. 1-28. Internal or external shorts and open circuits cause stuck-at trouble symptoms, or a dead section. *(Courtesy Hewlett-Packard Co.)*

highs and stuck lows, and to distinguish between single-shot responses, pulse streams, and low repetition highs and lows. As noted previously, a logic pulser permits injection of test pulses on one node, or on several nodes simultaneously.

Scanner-Monitor Digital-Logic Section Checks

With reference to Fig. 1-29, a basic digital-logic section for a scanner-monitor radio receiver employs a *clock* (free-running multivibrator), two *flip-flops* (bistable multivibrators), and four AND gates. Since the entire system is driven by clock pulses, preliminary tests are made without a pulser. Waveforms at all key test points can be checked with a logic probe. A square-wave train is normally present at the clock output. Half frequency square waves in opposite polarities are normally present at the flip-flop (FF) Q and not Q (Q̄) outputs (FF1 outputs). Quarter frequency square waves in opposite polarities are normally present at the FF2 Q and Q̄ outputs. Incorrect output, or absence of output at a test point indicates a fault in the associated device. Note that the array of square waves depicted in the diagram comprise the specified *timing diagram* for the digital system (see also Fig. 1-30).

> Note: All digital-logic circuits, no matter how complex, are direct coupled throughout. Switching transistors are used exclusively; input and outputs will always be effectively at ground potential or at power-supply potential (in normal operation).

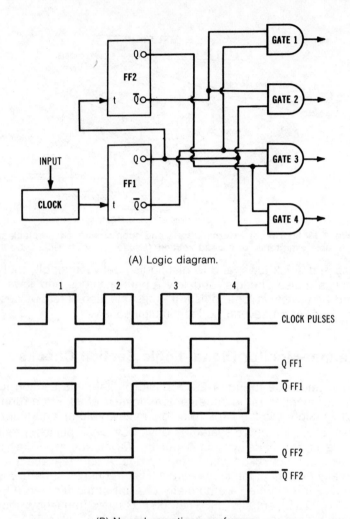

(A) Logic diagram.

(B) Normal operating waveforms.

Fig. 1-29. Simplified block diagram of a 4-channel digital-logic section in a scanner-monitor radio receiver.

Positive Logic and Negative Logic

Most digital systems operate with positive logic. This means that *low* is ground potential (or near ground potential). On the other hand, some digital systems operate with negative logic. This means that *high* is ground potential (or near ground potential). As depicted in Fig. 1-31, positive logic may be used with either a

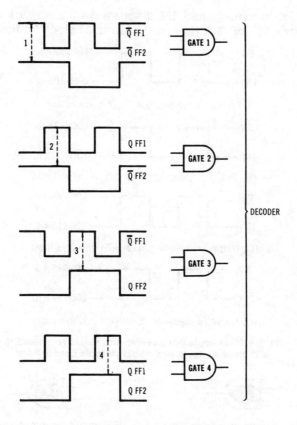

Fig. 1-30. Scanner-monitor AND gate inputs.

positive operating level or with a negative operating level. Similarly, negative logic may be used with either a positive operating level or with a negative operating level. We may encounter negative logic mixed with positive logic in a system to simplify the design. It follows from the truth tables in Fig. 1-19 that an AND gate operates like an OR gate if the AND gate is a negative logic circuit. Again, an OR gate operates like an AND gate in a negative logic circuit. Similarly, a NAND gate operates like a NOR gate in a negative logic circuit; a NOR gate operates like a NAND gate in a negative logic circuit. An XOR gate operates like an $\overline{XOR}$ gate, and vice versa, in a negative logic circuit (see Figs. 1-32 and 33).

Note: Any logic network can be discussed either in terms of positive logic, or of negative logic, just as any

43

electric circuit can be discussed in terms of conventional current flow, or in terms of electron flow.

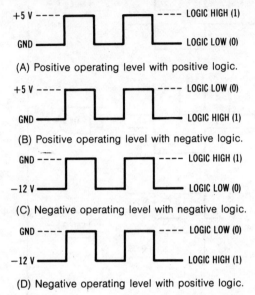

(A) Positive operating level with positive logic.

(B) Positive operating level with negative logic.

(C) Negative operating level with negative logic.

(D) Negative operating level with positive logic.

Fig. 1-31. Basic examples of positive and negative operating levels with positive logic and negative logic conventions.

(A) An AND gate in positive logic "looks like" an OR gate in negative logic.

(B) An OR gate in positive logic "looks like" an AND gate in negative logic.

(C) A NAND gate in positive logic "looks like" a NOR gate in negative logic.

(D) A NOR gate in positive logic "looks like" a NAND gate in negative logic.

(E) An XOR gate in positive logic "looks like" an XNOR gate in negative logic.

(F) An XNOR gate in positive logic "looks like" an XOR gate in negative logic.

Fig. 1-32. Gate equivalences in positive and negative logic systems.

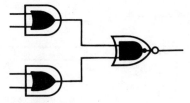

Fig. 1-33. An AND-OR invert arrangement in positive logic "looks like" an OR-AND-invert arrangement in negative logic.

Application of Logic Current Tracer

The more that a technician familiarizes himself with a failed digital circuit, the greater the likelihood that he or she can repair it quickly. There are many troubleshooting jobs wherein a faulty circuit node is identified as "stuck at," and if there are many elements common to this node, a "tough-dog" problem can result. This type of problem can be solved quickly and nondestructively by using a current tracer (see Fig. 1-5). It is no longer necessary to "lift" IC legs, to cut traces, or to force large amounts of dc current into the conductor in an attempt to burn out the short. A current tracer tip is held over a *pulsing* current path—it shows whether current is flowing. Sometimes no current flows because a node driver is dead. Most importantly, a current tracer shows *where* a pulsing current is flowing. If a node is stuck *low,* and the reason for the fault is a shorted input on one of the node components, a very strong pulsing current will flow between the node driver and the faulty component. If "crosstalk" is encountered, the tracer sensitivity is reduced to a chosen level which will eliminate or at least minimize crosstalk pickup.

Reference Settings for Current Tracer

It was previously mentioned that pc conductors (traces) that vary greatly in width cause flux density changes under a current tracer tip, as depicted in Fig. 1-34. This variation may be important when tracing supply-to-ground shorts, and the sensitivity of the current tracer may need to be changed slightly for optimum indication. Setting a reference on a node identified as faulty is fundamental in the current-tracing process, but the setting for that particular node has little, if any, relevance for other nodes, due to fan-out (branched loads) or variability in circuit interconnections. Also, the sensitivity control on the current tracer allows the technician to "see" currents as small as 300 μA, but there is virtually no upper limit. As Fig. 1-34 shows, if the sensitivity control is set so that 10 mA barely lights the tracer display, 30 mA will produce half-brilliance, and full brilliance is reached at 50 mA or slightly more. Currents greater than 50 mA *also* produce full brilliance.

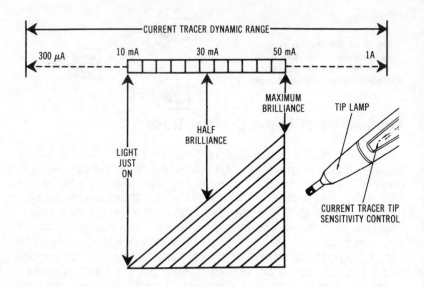

RANGE			
Settings	Current need for barely lit lamp	Current when lamp set for half-brilliance	Current that will produce a fully bright lamp
1	300 μA	1 mA	≥ 3 mA
2	1 mA	3 mA	≥ 5 mA
3	3 mA	5 mA	≥ 10 mA
4	5 mA	10 mA	≥ 30 mA
5	10 mA	30 mA	≥ 50 mA
6	30 mA	50 mA	≥ 100 mA
7	50 mA	100 mA	≥ 500 mA
8	300 mA	500 mA	≥ 1 A

When a 10 mA current transition occurs, and SENSITIVITY is set for half-brilliance of the tip lamp, the dynamic range of the current tracer is as follows:

DIM LAMP . . . . .5 mA
HALF-BRIGHT LAMP . . . . .10 mA
FULLY LIT LAMP. . . . .≥30 mA

The table above shows several examples of dynamic range for the full range of possible sensitivity control settings available.

Fig. 1-34. Sensitivity control settings for a current tracer.
(Courtesy Hewlett-Packard Co.)

Current-Tracer Operating Factors

Adjustment of the sensitivity control on a current tracer has been previously noted as a method of alleviating crosstalk interference. Also, the tracer tip is directional. That is, current paths oriented 90° out of phase with the pickup coil tend to null out. In turn, proper tip orientation (alignment) helps to eliminate crosstalk from conductors on different layers of a board or at different angles. Tracer orientation is illustrated in Fig. 1-35. In areas where a pc board has many conductors side by side carrying subtantial currents, the technician can move away from these areas to trace current from one component to another by setting a reference current level right on the node driver output pin. Then, the technician simply moves from pin to pin on the ICs, instead of attempting to follow along the conductors.

TRACER ORIENTATION
(Always position or orient Current Tracer for maximum coupling under the tip)

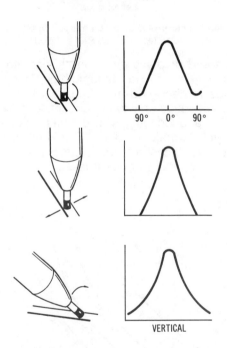

Fig. 1-35. Orientation (alignment) of current-tracer tip.
(Courtesy Hewlett-Packard Co.)

Current Tracer Senses AC Only

As depicted in Fig. 1-36, a logic current tracer is an ac device. It detects and displays current pulses or current transitions, and

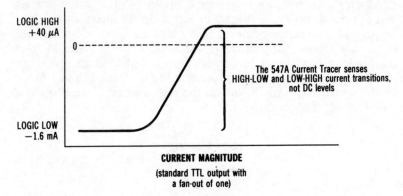

Fig. 1-36. A logic probe senses changes in current values.
(Courtesy Hewlett-Packard Co.)

then stretches and displays them via its indicator lamp. When a TTL output goes from logic low to logic high, for example, the total current change is about 1.6 mA. A current tracer is not voltage sensitive, and responds only to current changes. Note that CMOS, low-power TTL, and Schottky TTL circuits operate at somewhat lower current levels than standard TTL circuits, as shown in Fig. 1-37. In all cases, the short-circuit current level is typically 10 times the normal operating current level.

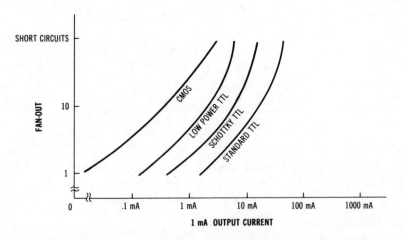

Fig. 1-37. Alternating current levels in normal and shorted digital circuits. *(Courtesy Hewlett-Packard Co.)*

Equivalent Gates in Digital Logic Circuits

The basic AND, NAND, OR, and NOR gates are easily recognized in digital logic diagrams. However, equivalent gates that perform the same functions are occasionally overlooked. For example, with reference to Fig. 1-38, a NOR gate has the same truth table as an AND gate with inverters connected in series with its inputs. Similarly, a NAND gate has the same truth table as an OR gate with inverters connected in series with its inputs. Again, an OR gate has the same truth table as a NAND gate with inverters connected in series with its inputs. It is evident that an AND gate has the same truth table as a NOR gate with inverters connected in series with its inputs.

A 2-input AND gate with an inverter connected in series with one of its inputs has the same truth table as a NOR gate with an

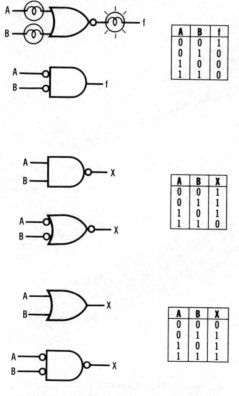

A	B	f
0	0	1
0	1	0
1	0	0
1	1	0

A	B	X
0	0	1
0	1	1
1	0	1
1	1	0

A	B	X
0	0	0
0	1	1
1	0	1
1	1	1

Fig. 1-38. Examples of equivalent gates.

inverter connected in series with one of its inputs. A 2-input NAND gate with an inverter connected in series with one of its inputs has the same truth table as an OR gate with an inverter connected in series with one of its inputs. Recognition of equivalent gates can be very helpful in checking out digital-logic circuitry.

Service-Type Logic Probes

Service-type logic probes (and pulsers), such as illustrated in Fig. 1-39, are comparatively economical and are widely used by digital technicians. The probe in this example has high and low LED indicators, and includes a pulse stretcher. A switch is provided for application in either TTL or CMOS circuitry. The companion pulser has one-shot (single pulse) output, or a manually controlled 5-Hz pulse train output. Another example of the service-type digital probe is shown in Fig. 1-40. This type of probe is designed for use in TTL, CMOS, or DTL circuitry. It provides

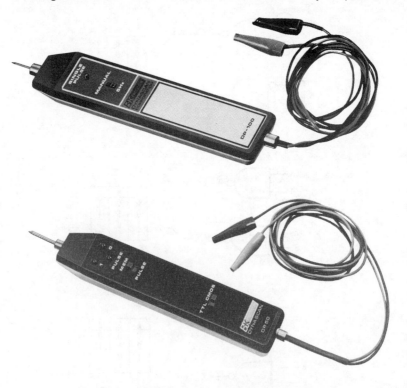

Fig. 1-39. Service type digital probe and digital pulser.
(Courtesy B & K Precision)

Fig. 1-40. A service type digital-logic probe.
(Courtesy Global Specialties Corp.)

high, low, and pulse LED indicators. A pulse stretcher is included.

Some digital technicians do not require high performance probes, and build their own, as shown in the example of Fig. 1-41.

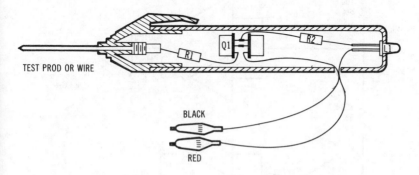

Parts List

Symbol	Qty	Description	Calectro Cat. No.
Q1,Q2	2	npn silicon transistor	K4-507
R1	1	27,000 ohm, ½ watt resistor	B1-401
R2	1	150 ohm, ½ watt resistor	B1-374
L1	1	LED Lamp	K4-559
—	1	GC Cat. No. 984 Pen Oiler	—
—	2	jumper wires	J4-650

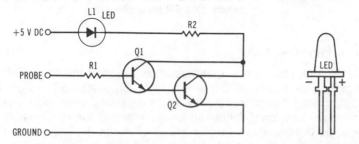

Fig. 1-41. Simple digital-logic test probe for checking input/output relations of gates or circuits. *(Courtesy Calectro)*

This arrangement is suitable for TTL circuit tests. It has a single LED indicator; if the test point is high, the LED glows. On the other hand, if the test point is low, the LED is dark. Next, the probe depicted in Fig. 1-42 provides two LEDs, for indication of high and low logic levels. It is designed for use in TTL circuitry. If the test point is low, one LED glows; if the test point is high, the other LED glows. In case that the point under test is open circuited, neither LED glows.

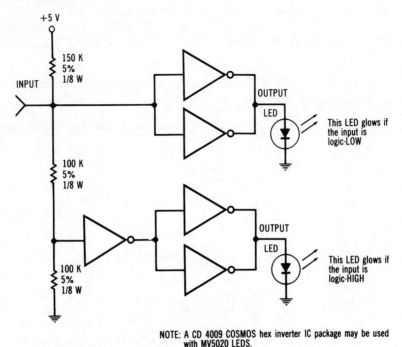

NOTE: A CD 4009 COSMOS hex inverter IC package may be used with MV5020 LEDS.

Fig. 1-42. Example of series-parallel inverting buffer circuitry used in a simple digital-logic probe.

Logic Flow Diagrams

Logic diagrams are sometimes shown in the form of logic-flow diagrams, as shown in the example of Fig. 1-43. When trouble-shooting logic circuitry, the technician usually does not have a logic flow diagram to follow. In turn, the technician must visualize the logic-flow relations. If a number of gates are involved, it is sometimes helpful to mark down the logic states on every node. Then, when a logic probe and pulser are used to check circuit operation, it is immediately apparent whether the circuit action is normal or abnormal. A specialized flow diagram with generalized logic equations is illustrated in Fig. 1-44. This configuration is explained in full detail in Section 2.

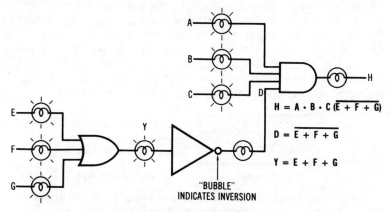

$$H = A \cdot B \cdot C \overline{(E + F + G)}$$

$$D = \overline{E + F + G}$$

$$Y = E + F + G$$

"BUBBLE" INDICATES INVERSION

Fig. 1-43. Logic flow diagram for an AND and OR gate application.

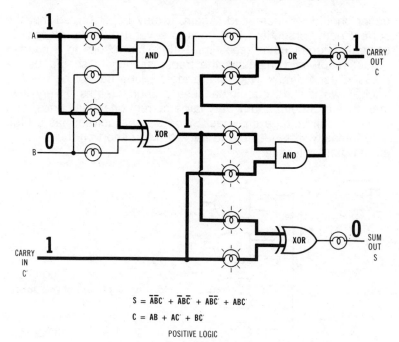

$$S = \overline{A}\overline{B}C' + \overline{A}B\overline{C'} + A\overline{B}\overline{C'} + ABC'$$

$$C = AB + AC' + BC'$$

POSITIVE LOGIC

Fig. 1-44. A specialized flow diagram and the generalized logic equations for a full adder.

Common Causes for Puzzling Observations

Sometimes a current path will seem to disappear as the current tracer is moved along the path. In such a case, it may be observed that the conductor becomes wider, resulting in current "fan-out" and lessening the field intensity under the tracer tip, as seen in Fig. 1-45. In turn, the sensitivity of the current tracer should be increased. Again, it may be observed that the conductor proceeds through a plated-through hole in the pc board; since the tracer is then farther from the current flow, the sensitivity of the

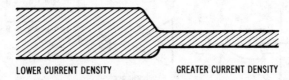

LOWER CURRENT DENSITY   GREATER CURRENT DENSITY

Fig. 1-45. Reason for current-density variation in pc conductors.
(Courtesy Hewlett-Packard Co.)

tracer should be increased. Again, it may be observed that the current path "branches" and goes to several different places via several different paths (see Fig. 1-46). This results in a sudden reduction of field intensity; the tracer sensitivity should be increased accordingly. Note that a logic probe cannot be used to track down a short circuit because a short creates a very low-impedance circuit. In turn, very large current changes correspond to very small voltage changes. Since a large current has a high field intensity, it is feasible to track down a short on the basis of current field intensity.

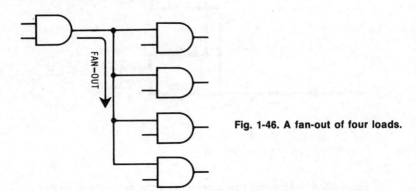

Fig. 1-46. A fan-out of four loads.

Sensitivity Control Settings

In typical digital troubleshooting situations, a current tracer is the final test instrument that will be used to pinpoint the fault on a node. Generally, the tracer pinpoints the fault by following current flow after voltage or logic-state sensing devices have been used to narrow the fault area down to a node, pc board conductor, or bus line (main digital information path). Then, since the area under test is probably stuck at a fixed voltage level, only the use of current tracing will indicate an activity path that can be tracked down. That is, a node "stuck" in one state may be trying very hard to change state, and in turn will be carrying a large amount of current. The sensitivity of a current tracer can be varied over a wide range of current values, from less than 1 mA to more than 1 A (see Fig. 1-47). In practical troubleshooting procedures, it is necessary to set the tracer sensitivity correctly, as follows:

1. Select the bad node, gate, or signal path to be traced.
2. Place the tip of the tracer at the node driver output.
3. Align the tip, as explained below.
4. Set the sensitivity control of the tracer for half brilliance on the indicator lamp.
5. Leave the sensitivity control at the same setting until the fault is located, or until test conditions change.

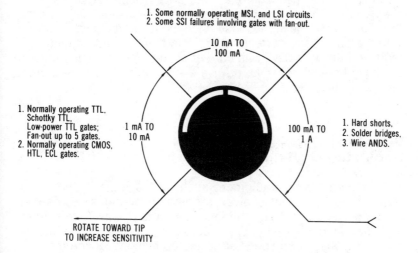

Fig. 1-47. Interpretations of current-tracer sensitivity settings.
(Courtesy Hewlett-Packard Co.)

The accompanying diagram (Fig. 1-47) can be useful to determine where a fault is located, or to identify the nature of the fault. For example, if a node is stuck and the sensitivity control of the tracer is set approximately midway (greater than 50 mA) during current tracer reference setting, it is likely that the node is good, and is driving current into a short circuit. The current-sensing coil of the tracer is very small (only 0.01 inch in diameter) although the protective plastic covering makes it appear larger. In turn, pc board conductors can be close together, and current tracing is still practical. For example, two conductors carrying identical current can be located 0.02 inch apart (edge to edge) and current flow between the two traces can be distinguished by the current tracer. Note, however, that if the two pc conductors (traces) have a 10-to-1 current difference between them; the minimum feasible spacing is 0.075 inch (see Fig. 1-48).

Digital IC Failure Modes

Digital ICs fail about three-fourths of the time by opening up at either the input or the output (see Fig. 1-49). These can be identified by logic probes and pulsers. Repairs to the other failures shown in the following diagram are facilitated by current tracing. A current tracer is most useful where the fault produces excessive current flow on a node. If there is little or no current flowing on the node, there is a likelihood of a dead driver (open output bond), or a lack of pulse activity in the circuit. In these situations, use the logic probe and pulser to close in on the fault area, and follow up with the pulser and current tracer. Note that V_{cc} and ground lines tend to be very "noisy" with *current* spikes. Avoid tracing current on these lines, if possible, or, use the pulser to boost the current on the node under test.

Waveform Checks in
Dedicated Digital Equipment

Dedicated digital equipment such as the logic section in a scanner-monitor radio receiver is operated by digital words that are automatically repeated indefinitely. In turn, waveform checks are made as described in Fig. 1-30; the displayed waveforms are compared with those specified in the equipment service manual. An oscilloscope suitable for digital-logic servicing procedures is illustrated in Fig. 1-50.

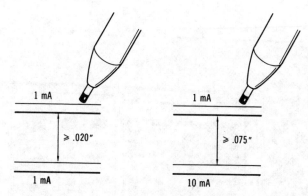

Fig. 1-48. Minimum feasible spacings between pc conductors for current-tracer tests. *(Courtesy Hewlett-Packard Co.)*

DIGITAL CIRCUIT FAILURE MODES

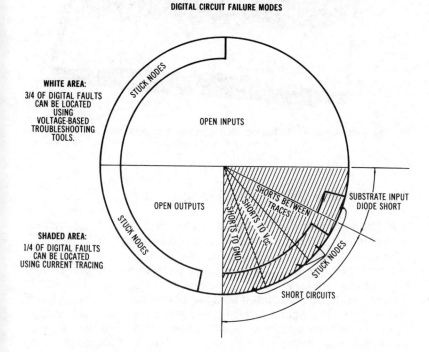

Fig. 1-49. Statistical summary of common faults in digital circuits.
(Courtesy Hewlett-Packard Co.)

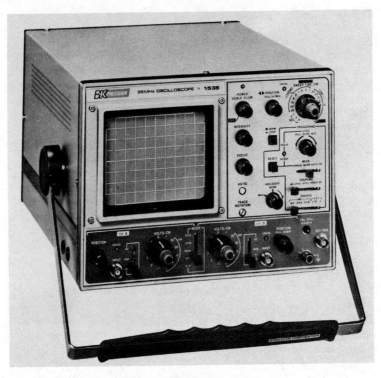

Fig. 1-50. Representative oscilloscope used in digital-logic servicing.
(Courtesy B & K Precision)

Digital Pulse Voltage Tolerance

Although digital ICs tend to fail catastrophically, situations will occasionally be encountered in which marginal defects result in attenuated pulse output. A gate is rated by the manufacturer for a minimum voltage level which is guaranteed to be interpreted as a high at the input, and for a maximum voltage level which is guaranteed to be interpreted as a low at the input. For example, in one widely used series of gates, 2 volts is the minimum voltage level that is guaranteed to be interpreted as a high at the input; 0.8 volt is the maximum voltage level that is guaranteed to be interpreted as a low at the input. The same series of gates is rated for a minimum output voltage in the high state of 2.4 volts; 0.4 volt is the rated maximum output voltage in the low state (see Fig.

1-51). Thus, it is good practice to observe the peak-to-peak voltage of any digital waveform, and to make certain that it is within tolerance. Otherwise, a good driven gate could be confused with a marginally defective driver gate. Similarly, it is good practice to check the power-supply voltage, and to make certain that it is not subnormal. When a power-supply defect causes V_{cc} to become marginal, gates that are near their tolerance limits will appear to be defective, although they will function normally when V_{cc} is brought back to its correct value.

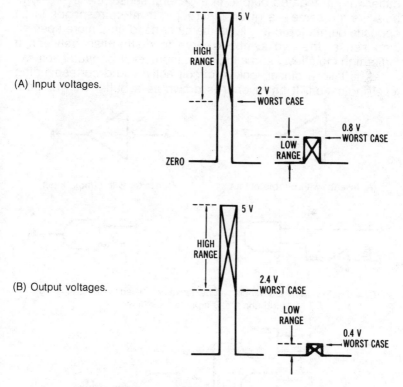

(A) Input voltages.

(B) Output voltages.

Fig. 1-51. Voltage tolerances in TTL digital systems.

Level Indicators

At this point, it is helpful to take note of level indicators used in logic diagrams. We have seen that if a buffer is followed by a small circle ("bubble"), that a logic-high input level will be changed into a logic-low output level, or vice versa. Accordingly, the small circle, or bubble, is a logic-level indicator. We say that the bubble *negates* the logic level (reverses the logic level). We will find that the *input* to a logic device, such as a gate or a buffer, may be negated. Let us review the logical meanings of negated inputs and negated outputs.

With reference to Fig. 1-52, a level indicator in a digital-logic configuration serves to indicate the logic level that will exist for the *intended function of the device.* For example, an inverter may be shown as a buffer with negated input, or it may be shown as a buffer with a negated output. In a general sense, these two symbols are the same—a given input logic level corresponds to an opposite output logic level. On the other hand, in a more specialized sense, the two symbols serve to distinguish between a logic-high input level and a logic-low input level *for circuit activity.*

Let us take a closer look at circuit activity and corresponding level indication. If an inverter is drawn as a buffer with a "bub-

(A) Inverter with bubble output.   (B) Inverter with bubble input.

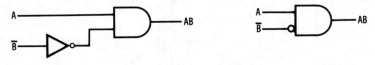

(C) AND gate with negated input will be active only when the A input is logic high and the B input is logic low.

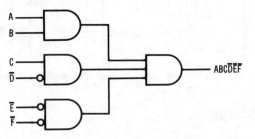

(D) Of the 64 possible input state combinations for this configuration, only the indicated combination will produce a logic-high output.

Fig. 1-52. Good practices for indication of logic levels.

bled" output, this means that the following logic device or circuit will be triggered (or will be otherwise active) when a logic-high state is inputted by the inverter. On the other hand, if an inverter is drawn as a buffer with negated (bubbled) input, this means that the following logic device or circuit will be active when a logic-low state is inputted by the inverter. In the first case, we say that the circuit is active-high; in the second case, we say that the circuit is active-low.

Consider next the AND gate with a negated input in Fig. 1-52. It is evident that the circuit will be active only when the A input is logic-high and the B input is logic-low. Therefore, the level indicators for the AND gate are written as A, B, and AB. The basic importance of appropriate logic-level indication is apparent in the diagram showing an AND gate, an AND gate with a negated input, and an AND gate with both inputs negated. This arrangement has 64 possible input state combinations! However, only one of these

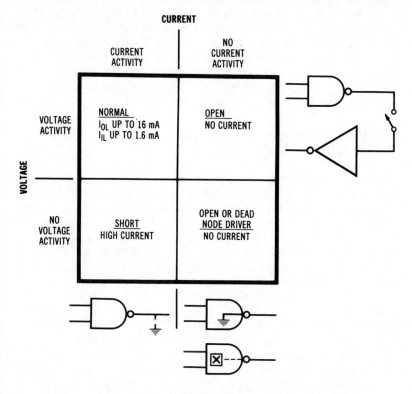

Fig. 1-53. Evaluation of voltage and current activity in failed ICs.
(Courtesy Hewlett-Packard Co.)

combinations will produce a logic-high output. This combination is $AB C\overline{D}\,\overline{E}\,\overline{F}$, and it is read A and B and C and not D and not E and not F.

It is clear that when level indicators are chosen to indicate the logic levels that will exist for the intended function of a device or circuit, that digital logic diagrams become much easier to read.

Current and Voltage in Failed ICs

A useful way of classifying IC failures is shown in Fig. 1-53. Note that in the upper left-hand corner, both voltage and current activity are present on a TTL circuit node in normal operation. Therefore, this is a preliminary evaluation factor. On the other hand, as noted in the lower right-hand corner, neither voltage nor current activity is present on a TTL node which is open or dead. At a short, there is no voltage activity but high current activity. At an open, there is no current activity but high voltage activity. Troubleshooting procedures may also be concerned with partial shorts which merely distort normal operation. For example, voltage activity can be subnormal with abnormal current activity; or, current activity can be subnormal with abnormal voltage activity.

In-Circuit Versus Out-of-Circuit Tests

Although an AND gate passes an out-of-circuit test, it should not then be assumed that it will operate normally in-circuit;—the gate will be required to drive a following device. The fan-out or drive capability of a TTL device denotes its ability to sink current in the output low state (see Fig. 1-54). In the output low state, the "phase splitter" transistor, Q2, is turned "on." It supplies base drive to the output pull-down transistor, Q4. The amount of base drive required for the pull-down transistor is dependent on the worst-case beta of the device and the fan-out (I_{OL}) sink-current requirements of the circuit. The output high drive current (I_{OH}) of the device is supplied from the pull-up transistor, Q3. When the phase-splitter transistor, Q2, is turned "off," the pull-up transistor is turned "on." This presents a low-impedance drive source at the output. Although the static I_{OH} requirements of most circuits are less than 0.5 mA, about 35 mA is made available at the instant of low to high output transition to charge up the distributed line, board, and package capacitances encountered in most system designs. Different types of pull-up circuits are used to achieve faster system speeds by minimizing high output impedance and the resulting rc time constant.

> Note: LOGIC PROBE, PULSER, AND CURRENT TRACER Versus OSCILLOSCOPE. Digital ICs fail about three-fourths of the time by opening up at either the input or output. This type of failure can often be found by using go/no-go voltage-based methods such as logic probes and pulsers. About one-fourth of digital trouble symptoms are caused by shorts to V_{cc} or to ground. This type of failure can often be found by using a go/no-go test with a current tracer. In the remaining trouble situations, electrical parameters such as peak-to-peak voltage, waveform timing relations, and noise peaks must be analyzed. As explained subsequently, glitch interference is occasionally an elusive troublemaker. These types of malfunctions are analyzed to best advantage with high-performance oscilloscopes.

Fan-In/Fan-Out Principles

It was noted previously that although a gate passes an out-of-circuit test, it should not then be assumed that the gate will oper-

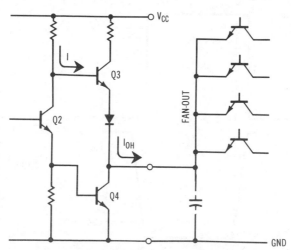

Fig. 1-54. Gate output circuit has limited drive (fan-out) capability.
(Courtesy Fairchild Camera and Instrument Corp.)

63

ate normally in-circuit; this depends on the ability of the gate to supply the drive current demand of the circuit. In order to simplify device specifications, it is customary to normalize the input and output loading parameters of all semiconductor families to the following values:

1 Unit TTL Load (U.L.)=40 μA in the high state (logic "1")
= 1.6 mA in the low state (logic "0")

Example: A Fairchild 9N00/7400 gate, which has a maximum I_{IL} of 1.6 mA and I_{IH} for 40 μA is specified as having an input load factor of 1 U.L. (Also called a fan-in of 1 load.)

Example: The 93H72, which has a value of I_{IL} = 3.2 mA and I_{IH} of 80 μA on the clock pulse terminal, is specified as having an input load factor of 3.2 mA divided by 1.6 mA, or 2 U.L.

Example: The output of the 9N7400 will sink 16 mA in the low (logic 0) state and source 800 μA in the high (logic 1) state. The normalized output low drive factor is, therefore, 16 mA divided by

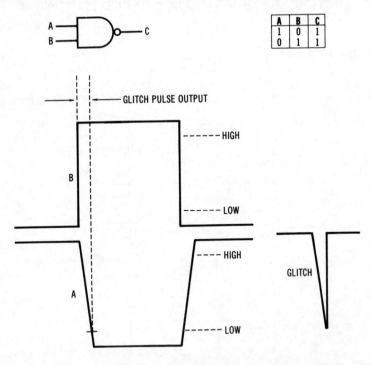

A	B	C
1	0	1
0	1	1

Fig. 1-55. NAND gate produces a race glitch when input B has fast rise and input A has slow fall.

1.6 mA, or 10′ U.L. The output high drive factor is 800 μA divided by 40 μA, or 20 U.L.

Glitches Caused by "Races" at Gate Inputs

Spurious pulses called glitches can have more than one cause; however, a common cause of very narrow glitches is a race condition at a gate. With reference to Fig. 1-55, the NAND gate initially has its A input high, and its B input low. Digital pulses are then applied simultaneously to the gate inputs. However, in this fault situation, the A waveform has a fall time that is longer than the rise time of the B waveform. Therefore, the A input remains high longer than in normal operation, with the result that a narrow glitch is generated and appears in the gate output waveform. To correct the malfunction, the troubleshooter must seek the cause of the abnormal fall time in the A waveform.

Noise Glitches in Digital Waveforms

Noise glitches, as depicted in Fig. 1-56, are usually of longer duration and more easily seen in a digital waveform than are race glitches. Noise voltages may have an internal source or an exter-

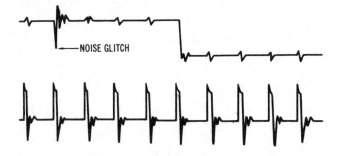

Fig. 1-56. Typical noise glitch in a digital waveform.
(Courtesy Hewlett-Packard Co.)

nal source. For example, although an electron travels very rapidly, in a large digital system, there are *propagation delays* from input to output which are easily measurable with a triggered-sweep scope. Propagation delays cause data changes to occur at slightly different times on data lines, resulting in internal noise. One line may pick up noise from another line. Noise voltages may also have an external source in industrial applications; marginal power supplies can also develop glitches. An oscilloscope can be utilized as a glitch-tracer, to track down a noise glitch to its source.

Section 2

ADDERS AND SUBTRACTERS

Combinatorial logic is defined as digital circuitry in which the states of the outputs from a device depend only on the states of the inputs. Gates and inverters are used in various arrangements to implement combination logic devices, and logic diagrams depict the interconnections of these gates and inverters.

Half Adder Operation

The use of an XOR gate with an AND gate in a half adder configuration was noted previously. When a scope check is made of half adder operation, four wave trains are involved, as seen in Fig. 2-1. Any pair of these waveforms may be simultaneously displayed on the screen of a dual-trace scope. Note in passing that the truth table in this diagram also applies to various other gate combinations that perform that half adder function. These alternative combinations follow from the principles outlined in Section 1.

Operation of Basic Full Adder

As seen in Fig. 2-2, the most basic full adder consists of two half adders with an OR gate. It will add two binary digits (bits) inputted at A and B. Also, if there is a carry in (C') from a previous adder, it will include the carry in. For example, if A, B, and C' are all driven high simultaneously, the sum Σ output will go high, and the carry out C output will also go high. In other words, $1+1+1 = 11$. In various applications, adders are operated in cascade, or in

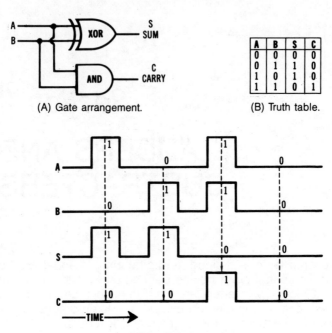

(A) Gate arrangement.

(B) Truth table.

A	B	S	C
0	0	0	0
0	1	1	0
1	0	1	0
1	1	0	1

(C) Operating waveforms.

Fig. 2-1. Half adder.

parallel. For example, two 2-bit adders are arranged in parallel so that pairs of binary digits can be added. Thus, 11+10 = 101. Still more adders may be included in a configuration to add larger binary numbers. We will find that adders are not only used for binary addition, but also for *binary subtraction, multiplication, division,* and so on. This is possible because the binary number system is simpler and more manipulative than the decimal number system.

Subtraction by Full Adders

With reference to Fig. 2-3, a 4-bit full adder will also operate as a subtracter when it is supplemented by four xor gates, Subtraction is accomplished by adding the 2's complement of the subtrahend to the minuend; the full adders then read out the difference. For addition, the "subtract" (carry in) terminal is held low. In turn, the B bits pass through the xor gates without inversion, and the full adders produce the sum of the input bits. For example,

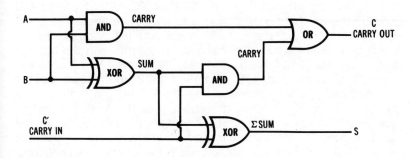

A	B	C′	S	C
0	0	0	0	0
0	0	1	1	0
0	1	0	1	0
0	1	1	0	1
1	0	0	1	0
1	0	1	0	1
1	1	0	0	1
1	1	1	1	1

Fig. 2-2. Most basic form of full adder sums two binary digits and includes a carry in (if present).

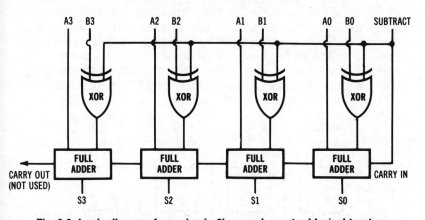

Fig. 2-3. Logic diagram for a simple 2's complement adder/subtracter.

suppose that the A bits are 1000, and that the B bits are 0011; in turn, the sum output is 1011 (in decimal terms, $8+3 = 11$). Next, suppose that 0011 is to be subtracted from 1000; the subtract terminal is driven high, whereby a 1 is entered into the first full adder. The minuend 1000 enters the adders without any processing. However, the subtrahend 0011 becomes complemented through the inverters (XOR gates), and enters the adders as 1100. This processed subtrahend, 1100, is called the 1's complement of the B bits. Since 1 has previously been entered into the first full adder, the 1's complement effectively becomes 1101; this further processed subtrahend, 1101, is called the 2's complement of the B bits. Finally, the full adders sum the A bits, and the 2's complement of the B bits give a readout of 0101. In this summing operation, the carry out is cast out (not used), as required by the rules (algorithm) for subtraction by 2's complement. In decimal terms, $8-3 = 5$. The XOR gates are said to operate as *controlled inverters*. Use of a 2's complement adder/subtracter simplifies the configuration for an arithmetic-logic unit (ALU), which would otherwise require a full subtracter section to supplement the full adder section. Note in passing that multiplication is a form of repeated addition; for example, $3\times5 = 5+5+5 = 3+3+3+3+3$.

Parallel Adder Function

To speed up addition operations, many high-speed digital systems use *parallel addition*. The time required in a parallel addition is dependent on the *propagation delay* of the carry bit. *Ripple carry* denotes a basic process of addition wherein the carries flow through to the output after the individual inputs have been summed. This limiting factor in operating speed is overcome by using carry look-ahead circuitry; the digital troubleshooter occasionally needs to be aware of propagation delay due to ripple carry. A block diagram for a 4-bit parallel adder is shown in Fig. 2-4. Observe that the data must be available in parallel at the

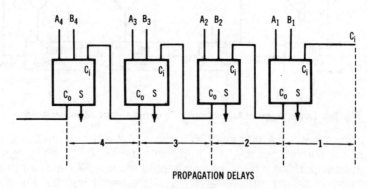

PROPAGATION DELAYS

Fig. 2-4. Block diagram of a 4-bit parallel adder.

input of the adder. After the necessary time to allow the output to "settle," due to the propagation delay of the carry bit, it is entered in parallel to the adder output register.

Sign-Magnitude Adder/Subtracter

Sign and magnitude adder/subtracters are often encountered, particularly in digital equipment that has been in service for some time. This type of adder/subtracter generally utilizes 1's complements of the binary numbers (Fig. 2-5), whereas the controlled-inverter adder/subtracter noted above employs the 2's complements of the binary numbers.

A typical sign and magnitude adder/subtracter is shown in Fig. 2-6. When this method of addition and subtraction is used, both positive and negative numbers are stored in uncomplemented form in the computer memory. The distinction between positive numbers and negative numbers is their sign bit. Note that unlike 1's complement and 2's complement arithmetic, addition and subtraction operations in the sign and magnitude format are not the same for all combinations of numbers. To determine which (if any) of the numbers must be complemented in an addition or

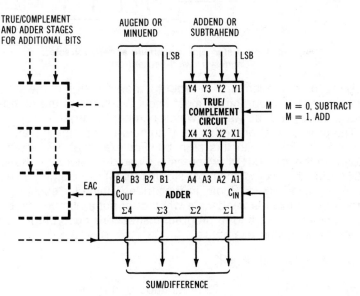

Fig. 2-5. Typical 1's complement adder/subtracter.
(Courtesy Hewlett-Packard Co.)

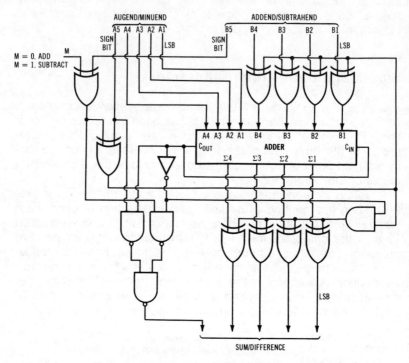

Fig. 2-6. Typical sign and magnitude adder/subtracter.
(Courtesy Hewlett-Packard Co.)

subtraction operation, and in order to know the sign of the sum or difference, the programmer must know the signs of the numbers that are involved, as well as which number is larger.

The programmer precedes each number with a sign bit, and the number is stored with this sign bit. For example, +7 would be stored in the form 00111, whereas −7 would be stored in the form 10111. In other words, the most-significant bit is the sign bit and is so recognized by the logic in the adder/subtracter section. The programmer must also observe whether or not a number needs to be complemented, and to program instructions accordingly. Consider the following examples:

$$+13 + 10 = +23$$

Augend	001101	To magnitude of augend,
Addend	001010	add magnitude of addend.
Sum	010111	Retain existing sign (0 for +).

$$-13 - 10 = -23$$

Augend	101101	To magnitude of augend,
Addend	101010	add magnitude of addend.
Sum	110111	Retain existing sign (1 for −).

Observe that the most-significant bits (sign bits) are not added in the foregoing example—the negative sign bit was merely retained in the sum.

$$-13 + 10 = -3$$

Augend	101101
Addend	001010

Here, the programmer instructs the CPU to add the 1's complement of the addend magnitude to the augend magnitude:

Magnitude of augend	01101	
1's complement of addend magnitude	10101	
	00010	
	1	End-around carry (q.v.)
Magnitude of sum	00011	
Final answer	100011	Retain sign of augend (1 for −).

$$+13 - 10 = +3$$

Augend	001101
Addend	101010

Here again, the programmer instructs the CPU to add the 1's complement of the addend magnitude to the augend magnitude:

Magnitude of augend	01101	
1's complement of addend magnitude	10101	
	00010	
	1	End-around carry
Magnitude of sum	00011	
Final answer	000011	Retain sign of augend (0 for +).

Note that all of the foregoing examples are defined to be additions. Subtraction requires the complementing of the subtrahend in a true/complement circuit, as shown in Fig. 2-5. This is a 1's complement adder/subtracter configuration. When this arrangement is used, the mode line M is driven logic-low for subtraction; conversely, the mode line is driven logic-high for addition. Observe that the programmer must frame a subtraction problem as an addition problem. For example, +13+10 can also be processed as +13−(−11).

Next, when the 2's complement form of addition and subtraction is utilized, positive binary numbers are stored in the computer memory in binary code, with a 0 in the sign position. On the other hand, negative numbers are stored in 2's complement form; a sign bit of 1 indicates that the sign is negative. Addition is performed directly; subtraction requires complementing of the subtrahend before addition is performed. As noted above, the programmer must frame a subtraction problem as an addition problem; thus, +13+10 can also be processed as +13−(−11). Observe that unlike operations with 1's complement numbers, 2's complement operations never generate an end-around carry. Consider the following examples of the 2's complement method:

$$+13 + 10 = +23$$

Augend	001101	To augend magnitude
Addend	001010	add addend magnitude.
Answer	010111	Retain existing sign.

$$-13 - 10 = -23$$

Augend	110011	To 2's complement of augend magni-
Addend	110110	tude, add 2's complement of addend
		magnitude, and neglect the last carry.
Answer	001001	Retain existing sign.

Note that the answer in the foregoing example is in 2's complement form. To change the answer into binary-number form, subtract 1 from the answer and then change all 1's to 0's and all 0's to 1's.

The foregoing problems involved two numbers of the same sign. Next, consider two numbers of the opposite sign, wherein the augend magnitude is larger:

$$+13 - 10 = +3$$

Augend	001101	To magnitude of augend add 2's com-
Addend	110110	plement of addend; also add sign bits,
Answer	000011	but neglect carry from sign bits.

$$-13 + 10 = -3$$

Augend	110011	To 2's complement of augend
Addend	001010	magnitude add magnitude of addend;
Answer	111101	also add sign bits, but
		neglect carry from sign bits.

Note that the answer is in 2's complement form. To change the answer into binary-number form, subtract 1 from the answer and then change all 1's to 0's and all 0's to 1's.

The foregoing problems involved two numbers of opposite sign, with the augend magnitude larger than the addend magnitude. Next, consider two numbers of opposite sign, wherein the addend magnitude is larger (or equal):

$$-10 + 13 = +3$$

Augend	110110	To 2's complement of augend
Addend	001101	magnitude, add magnitude of addend;
		add sign bits also,
Answer	000011	but neglect carry from sign bits.

$$+10 - 13 = -3$$

Augend	001010	To magnitude of augend add 2's com-
Addend	110011	plement of addend; also add
		sign bits, but neglect
Answer	111101	carry from sign bits.

Note that the answer is in 2's complement form. To change the answer into binary-number form, subtract 1 from the answer and then change all 1's to 0's and all 0's to 1's.

The basic serial adder arrangement is depicted in Fig. 2-7. A serial adder for more than two binary numbers is configured around a single full adder, as shown in Fig. 2-8. The auxiliary logic that is employed is arranged so that the augend word is entered from an external memory into the A shift register. The addend is entered into the B shift register. Both augend and addend have their least-significant bits on the right. Addition starts with the shifting of the least-significant bit from each register into the full adder. (See Fig. 2-9). Then, the sum is shifted into the

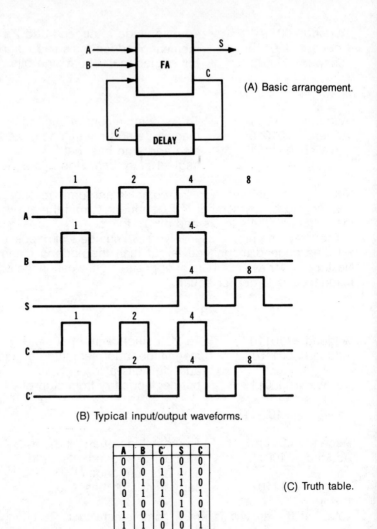

(A) Basic arrangement.

(B) Typical input/output waveforms.

A	B	C'	S	C
0	0	0	0	0
0	0	1	1	0
0	1	0	1	0
0	1	1	0	1
1	0	0	1	0
1	0	1	0	1
1	1	0	0	1
1	1	1	1	1

(C) Truth table.

Fig. 2-7. Serial adder.

sum register, and the next two bits from the A and B registers are shifted into the adder on the same clock pulse.

If the first addition produces a carry, this carry is stored for one clock pulse in the carry storage flip-flop; it becomes a carry in to the full adder on the next clock pulse. Note that a pair of binary words of any length can be added in this manner, starting with the

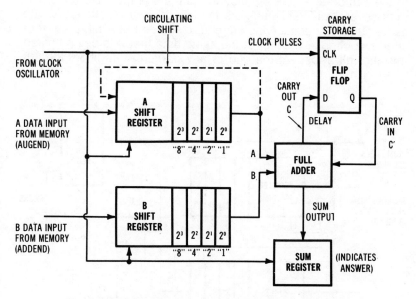

Fig. 2-8. Serial adder for more than two binary numbers.
(Courtesy Hewlett-Packard Co.)

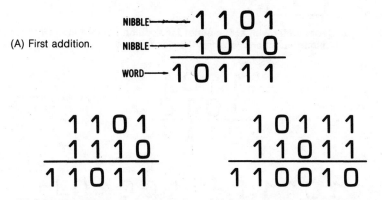

(A) First addition.

NIBBLE——**1 1 0 1**
NIBBLE——**1 0 1 0**
WORD——**1 0 1 1 1**

1 1 0 1
1 1 1 0
1 1 0 1 1

(B) Second addition with the original augend.

1 0 1 1 1
1 1 0 1 1
1 1 0 0 1 0

(C) Sum register indicates total sum of the two additions.

Fig. 2-9. Example of serial addition for more than two binary numbers.

least-significant bit and ending with the most-significant bit, and clocking all of the shift registers with the same clock signal. Next, if it is desired to add more than two binary numbers in sequence to the same augend, a circulating shift line is included. This permits several different numbers to be added in sequence to the same augend, because the augend is shifted back into the A register in a circular manner at the same time that it is being shifted into the full adder. At the end of any particular addition, the au-

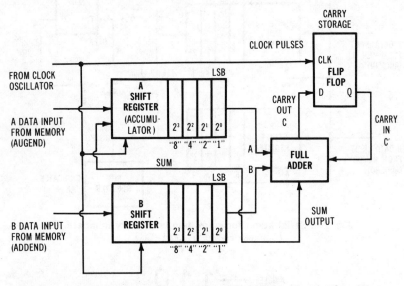

Fig. 2-10. Serial adder with accumulator for addition of more than two binary numbers. *(Courtesy Hewlett-Packard Co.)*

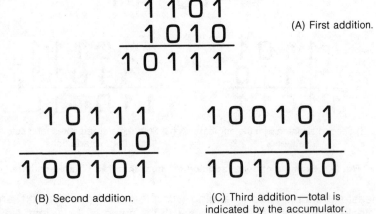

Fig. 2-11. Example of serial addition with serial adder and accumulator.

gend is back in its original place in the A register, ready for another addition. (See also Recirculating Memory.)

An alternative configuration for a serial adder that computes the sum of more than two numbers is shown in Fig. 2-10. The first two numbers are added together; the third number is added to the sum of the first two, and so on (see Fig. 2-11). This is accomplished by shifting the sum back into register A; the register is called an accumulator in this configuration.

Section 3
LATCHES AND FLIP-FLOPS

A latch is a bistable multivibrator used for temporary storage of binary information in a digital system. It is a temporary *memory* that can be erased, or *reset,* as required. It stores data for a specified time, and then unloads it. Two NOR gates are used in the simplest configuration (see Fig. 3-1). This is called an *RS* (reset-set) *flip-flop* or *RS latch;* it is the most basic form of flip-flop. At this time, we will use the terms *flip-flop* and *latch* interchangeably. Observe that cross-connection of the NOR gates provides positive feedback with resulting latch action. If the S (set) input is driven high, while the R (reset) input is held low, the $\overline{Q}$ output goes low; and Q must go high. The circuit remains latched until the R input is driven high. Thereupon, Q goes low, and $\overline{Q}$ must go high; the outputs thereby unload the stored information into the following circuit (not shown).

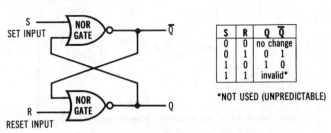

S	R	Q	$\overline{Q}$
0	0	no change	
0	1	0	1
1	0	1	0
1	1	invalid*	

*NOT USED (UNPREDICTABLE)

Fig. 3-1. Basic reset-set (RS) latch.

Feed-Through Latch

At this point, it is instructive to briefly consider the feed-through type of latch, which will be encountered on occasion in various digital logic diagrams. A feed-through latch, also known as fall-through latch, is depicted in Fig. 3-2. "Address" is synonymous with data in the present discussion. When the address latch-control signal is high, the address data "falls through" the latch. Next, when the address latch-control signal is low, the address data is latched into the circuit.

Observe in Fig. 3-2 that when the address latch-control signal is high, AND gate A1 is enabled, and AND gate A2 is disabled. Accordingly, data falls through the latch. Again, when the address

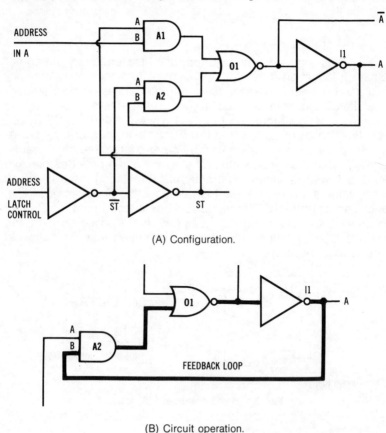

(A) Configuration.

(B) Circuit operation.

Fig. 3-2. Configuration of a feed-through latch.

latch-control signal goes low, AND gate A1 is disabled, thereby blocking any new data inputs. AND gate A2 is enabled by a logic-high on input A, which allows the feedback state to determine the input of gate A2. If the feedback state is low, the output of A2 will be low. On the other hand, if the feedback state is high, the output of A2 will be high. Note that there are two inversions from the output of gate A2 (O_1 and I_1) to the feedback loop. Consequently, the feedback action maintains the level that was present on inverter I_1 when the address latch-control state goes low.

Operation of a Delay (D) Latch

A basic D latch configuration is shown in Fig. 3-3. Because an inverter is used in series with the R input, complementary inputs are applied to the NOR gates, and race conditions are avoided. The D latch provides a simple solution to the race problem of the RS latch, and is a widely used circuit. The D latch has only one data input, and the latch is set ($Q = 1$) by clocking in a logic 1 and reset ($Q = 0$) by clocking in a logic 0. In operation, the clock signal is raised to logic high, and the desired high or low input is

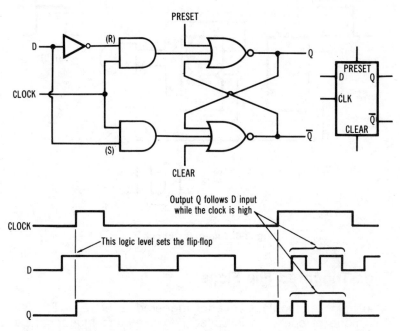

Fig. 3-3. Basic configuration of a D latch. *(Courtesy Hewlett-Packard Co.)*

applied to the data line; then the clock signal is allowed to go low before the data line input signal is changed. As soon as the clock goes low, the circuit is latched and the data input line may change arbitrarily without affecting the outputs. Alternatively, the D input can be applied prior to raising the clock signal high.

One-Pulse Circuit

Older designs of digital systems utilized monostable multivibrators to a greater extent than later designs. The one-pulse circuit, shown in Fig. 3-4, is preferred by many system designers because it is a synchronized device. The one-pulse circuit employs a pair of D flip-flops and an AND gate to generate an output pulse that has twice the width of the clock pulse. Note that the width of the output pulse is the same, whether the input of the circuit is triggered by a very narrow pulse, or by a wide pulse. Thus, its circuit action is analogous to that of a one-shot multivibrator.

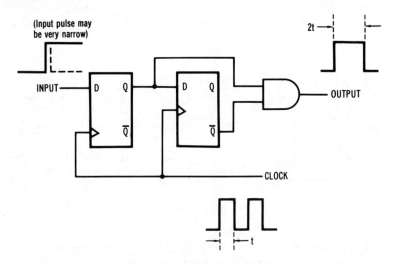

Fig. 3-4. One-pulse circuit.

Edge-Triggered Flip-Flops

Most clocked flip-flops are edge triggered, as shown in Fig. 3-5. This mode of operation is called negative-edge triggering, because data can be accepted from the J and K inputs only for a

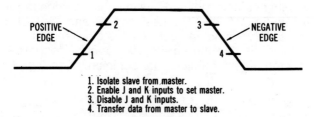

1. Isolate slave from master.
2. Enable J and K inputs to set master.
3. Disable J and K inputs.
4. Transfer data from master to slave.

Fig. 3-5. Timing relations along the clock pulse for a master-slave JK flip-flop.

brief interval prior to point 3 on the clock pulse. At 3, the J and K inputs are disabled, and remain disabled until the following positive edge rises to point 2. Although data on the J and K lines may change during the interval from 2 to 3, only the state of the data just prior to 3 will affect the following state of the slave. A small triangle is often indicated at the clock terminal to denote edge triggering, as seen in Fig. 3-4. This denotes positive edge triggering. If a small circle ("bubble") is included, negative edge triggering is denoted, as shown in Fig. 3-6. Fig. 3-6A shows a generalized symbol for a JK flip-flop. A high J produces a set on the next positive clock edge. A high K produces a reset on the next positive clock edge. If J and K are both high, there will be one toggle per positive clock edge. Fig. 3-6B shows a standard symbol for a JK flip-flop with preset and clear functions. Positive edge triggering is indicated. A high PR presets the FF, and a high CLR clears

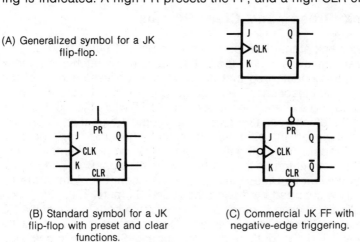

(A) Generalized symbol for a JK flip-flop.

(B) Standard symbol for a JK flip-flop with preset and clear functions.

(C) Commercial JK FF with negative-edge triggering.

Fig. 3-6. Edge-triggered symbols for JK flip-flops.

the FF. Another commercial JK FF with negative edge triggering is shown in Fig. 3-6C. A low PR presets the FF, and a low CLR clears the FF. If a triangle is not indicated at the clock terminal, the general conclusion is that the flip-flop is not edge triggered, but is level triggered. These relations are summarized in Fig. 3-7.

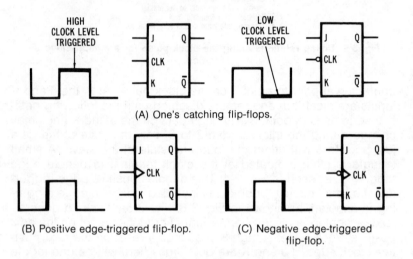

(A) One's catching flip-flops.

(B) Positive edge-triggered flip-flop.

(C) Negative edge-triggered flip-flop.

Fig. 3-7. Level-triggered and edge-triggered flip-flop symbols.

Clock, Preset, and Clear Signals

The clock signal is a particularly important pulse train. In Fig. 3-8, two gates are connected to the inputs of the latch and a clock signal is connected so that it can *enable* or *disable* both of the gates simultaneously. These gates prevent the R and S inputs from causing a change in the state of the FF while the clock is low. When the clock goes to a logic 1, any logic 1 signal on the R or S inputs is gated in (permitted to enter). Then the clock goes to logic 0 again, disabling the input gates and preventing other signals from entering the latch. The clock signal, in effect, produces a *window*. Unless this window is open, the state of the FF cannot be changed by driving the R and S inputs. Thus, the clock signal can be used to clock or gate data into both inputs of the RS latch. Preset and clear inputs are used to set or reset an FF without involving the data and clock inputs. Since the clock synchronizes a digital system, it is said that the preset and clear inputs are used to set and reset the FF asynchronously.

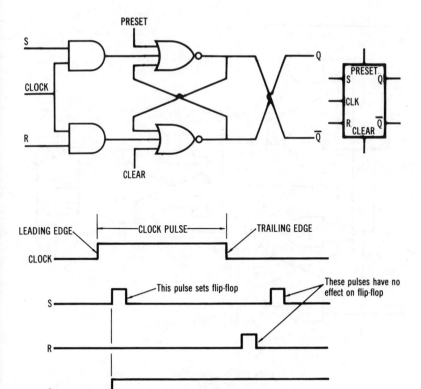

Fig. 3-8. Operation of a clocked RS latch. *(Courtesy Hewlett-Packard Co.)*

Operation of a D Flip-Flop

As seen in Fig. 3-9, a D flip-flop appears to be more complicated than a D latch. However, the two are closely related functionally. They both have a single data input and a logic 1 or logic 0 signal is used to set or reset them. The basic difference between the D flip-flop and the D latch is in the way that the clock signal is used for gating-in the data. Any change in the status of the output latch in the FF can take place only at the instant when the clock signal changes from a logic low to a logic high level—at no other time can an output change occur. The D flip-flop samples data present at the input only with the *rising edge* of the clock pulse, as shown in the timing diagram. This is known as *edge triggering,* and it is characteristic of most FFs, such as the D flip-flop and the JK flip-flop. The D flip-flop circuit consists of two interconnected

87

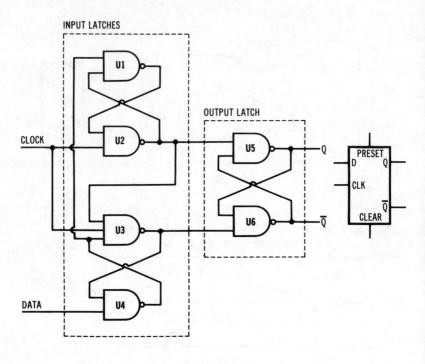

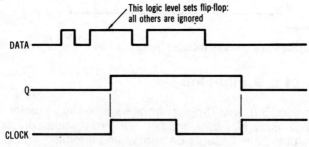

Fig. 3-9. D flip-flop configuration and timing diagram.
(Courtesy Hewlett-Packard Co.)

latches, gates U1/U2 and U3/U4, and an output latch, gates U5/U6. The input latches are interconnected so that when the clock goes from low to high (leading edge), it causes the input latches to lock in complementary states.

In other words, one input latch always supplies a logic 1 to the output latch, and the other input latch always supplies a logic 0 to the output latch. Which way they latch is determined by the state of the data line at the leading edge of the clock pulse. Once the clock goes high, it holds both input latches in their existing states and the data line can have no further effect. When the clock goes low again, both input latches supply a logic 1 to the output latch, and the data line can only affect the status of gates U1 and U4.

Operation of a JK Flip-Flop

A JK flip-flop, exemplified in Fig. 3-10, is similar to an RS latch in that it has two data inputs. However, the JK FF is advantageous in that it cannot have an undefined output, and its latches are not subject to the race condition. Like the D flip-flop, the JK flip-flop is clocked and is always edge triggered. However, most designs of the JK FF are controlled by the trailing edge instead of the leading edge. As noted previously, if both inputs of a JK FF are held at logic 1, the FF will change states when the clock edge occurs. That is, if it was set before the clock pulse, it will be reset after, and vice versa; this action is called *toggling*. The time before the clock edge is customarily called t, and the time after is called t_{n+1}. Similarly, the state of the Q output before the clock edge is termed Q_n and the state after as Q_{n+1}.

Flip-Flop Family Relations

As shown in Fig. 3-11, the RS flip-flop is the start of the flip-flop family group. The D flip-flop and the JK flip-flop are derived from the basic RS configuration. Also, the T flip-flop is obtained by feedback wiring of the RS flip-flop. Note also that the one-pulse circuit is a derivative of the RS configuration, as is the feed-through (fall-through) latch. The fall-through latch is also called a transparent latch.

Logic Probe, Pulser, and Current Tracer In JK Flip-Flop Troubleshooting

A flip-flop malfunction is generally investigated to best advantage with a logic probe, logic pulser, and current tracer, as depicted in Fig. 3-12. The logic probe serves to show which nodes (if any) are active, and the pulser provides signal injection at selected nodes. If the malfunction is due to a short, the current tracer is the practical instrument to localize the fault.

Case History*

In Fig. 3-12, the node between U1 and U2 was found to be stuck low when checked with a logic probe, although the probe showed pulse activity at the input of U1. Next, pin 2 was pulsed to see if the state of the node could be changed. In this example, the state couldn't be changed. Hence, the pulser and current tracer were utilized and it was found that input pin 9 on the JK flip-flop was shorted.

Another example of logic test-equipment application in a malfunctioning D flip-flop circuit is shown in Fig. 3-13. The logic probe and logic pulser are used to check input/output response.

Case History*

Two D flip-flops are shown in Fig. 3-13. One operates normally; the other doesn't change state although input conditions are identical for both. First, the pulser is applied at the D input, and the probe is applied at Q and $\overline{Q}$ to see if the outputs change. (In this case, they didn't.) Next, the reset line was found to be stuck low in a probe test; the reset line could not be driven high, indicating that the line was shorted to ground. Then the current tracer and pulser showed that the area near the reset line drew current when pulsed. Finally, a hairline solder bridge was found from the reset line to ground. When it appeared that the reset line was stuck low (shorted to ground), the current tracer and logic pulser were used to localize the short. Note that in this case, a normally operating D flip-flop was available for cross-checking of circuit activity.

---

*Courtesy Hewlett-Packard Co.

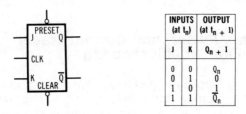

INPUTS (at t_n)		OUTPUT (at t_{n+1})
J	K	Q_{n+1}
0	0	Q_n
0	1	0
1	0	1
1	1	$\overline{Q}_n$

(A) Logic symbol and truth table.

Fig. 3-10. JK flip-flop

90

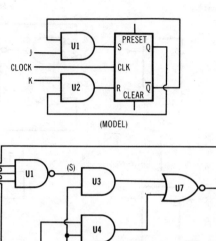

(MODEL)

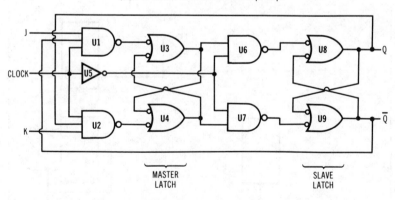

(B) RS latch version of JK flip-flop.

MASTER
LATCH

SLAVE
LATCH

(C) Master-slave version.

configurations.

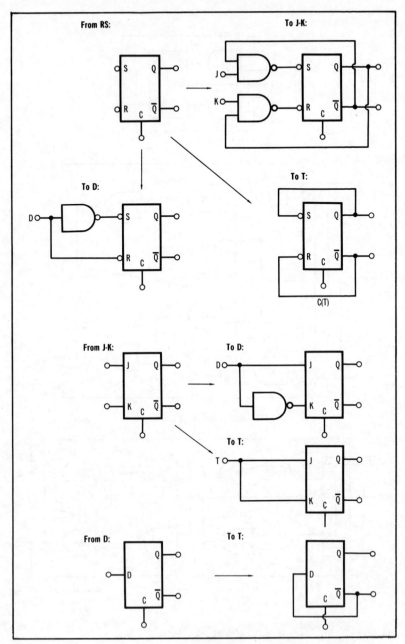

Fig. 3-11. Flip-flop family relations.

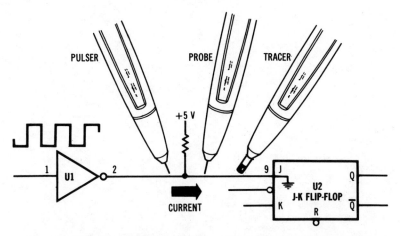

Fig. 3-12. Typical application of logic probe, pulser, and current tracer in flip-flop troubleshooting. *(Courtesy Hewlett-Packard Co.)*

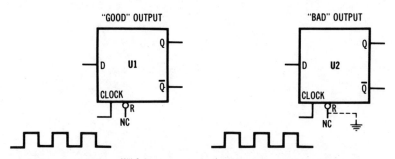

Fig. 3-13. Common application of logic probe, pulser, and current tracer in flip-flop troubleshooting. *(Courtesy Hewlett-Packard Co.)*

Section 4
SHIFT REGISTERS

With reference to Fig. 4-1, a shift register consists of a chain of flip-flops, interconnected as shown; the output from one flip-flop becomes the input of the next flip-flop. All of the flip-flops have a common clock signal, and all are set or reset at the same time. A logic 1 is applied to the D input of the first flip-flop, just long enough to be stored by the edge of the clock pulse. On the next clock pulse, FF2 will receive the 1 from the output of FF1; thereby, FF2 is set and the pulse is stored. Meanwhile, FF1 will sample its input again at the edge of the next clock pulse; if there is a logic 0 at its input, it will load a logic 0. On the third clock pulse, the 1 bit

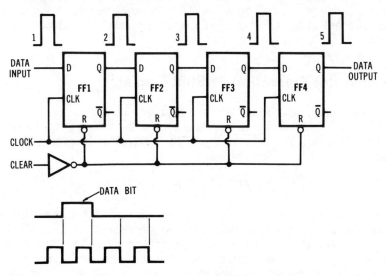

Fig. 4-1. Basic 4-bit shift register configuration.

will move from FF2 into FF3. On the fourth clock pulse, the 1 bit will move from FF3 into FF4. On the fifth clock pulse, the 1 bit will be unloaded from the Q and $\overline{Q}$ output terminals of the shift register.

When the clear input is pulsed, the contents of the shift register are erased, and all Q outputs go to 0 and all $\overline{Q}$ outputs go to 1. The clear action takes place independently of the clock pulse.

Data Storage in Shift Register

It is evident that a logic 1 is simply *shifted* or *stepped* through the FFs in a shift register by the clock signal. In turn, this single logic bit is regarded as an entity as it moves through the register; it is termed a *bit* or a *data bit*. Similarly, logic 0s in a shift register are called bits, and their logic state is just as important as the 1s in defining the contents of the register. In a 4-bit register, if more than four clock pulses are applied after a 1 is entered, the logic 1 will be shifted out of FF4 on the fifth clock pulse (unloaded), and all of the FFs will have logic 0 output levels again. A shift register may be used to store a digital number. If each bit in the register represents 1, the combined number of 1s will represent the stored number. Thus, if the register contains two 1 bits, the number 2 is stored; to store the number 3, three 1 bits must be contained in the four FFs. To enter the number to be stored in the register, four clock pulses must be applied and the appropriate logic level must be present at the D input of FF1 when the leading edge of each clock pulse arrives.

Converter Operation of Shift Register

With reference to Fig. 4-2, if the shift register is first cleared (all FFs reset) and the four data bits shown in the timing diagram are shifted through the register, the FFs will have the following states at the end of four clock pulses:

$$FF1 = 0 \quad FF2 = 1 \quad FF3 = 1 \quad FF4 = 1$$

This condition can be written simply as 0111. If, instead, the number 2 had been stored, it would be represented by 0011; again, the number 4 would be represented by 1111. With these four FFs, five numbers can be stored—0, 1, 2, 3, and 4. Because the bits were loaded one after another and the shift register shifted them from one FF to another, this sequence is called loading

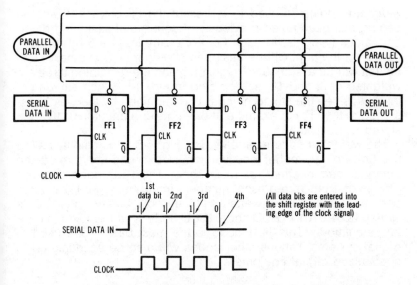

Fig. 4-2. Methods of shifting data into and out of a shift register.
(Courtesy Hewlett-Packard Co.)

serial data and the circuit is termed a *4-bit serially loaded shift register.* As has been observed, the alternative to serial loading of the shift register is *parallel loading.* In parallel loading, a separate line is connected to the preset input of each of the four FFs as shown in the diagram. All of the data bits are loaded simultaneously by setting the appropriate FFs to logic 1 through the preset input. This method of loading takes place without the use of a synchronizing clock signal; it is said to be *asynchronous.* The data on the four preset lines is called *parallel data.* Data bits can be loaded into the shift register in a parallel or a serial manner, and they can be read out all at once, or one at a time. A parallel readout is accomplished by simultaneously sampling the data at the output of each of the FFs. A serial readout is performed by shifting data through the FFs and sampling at the output of FF4. If the data is loaded serially and read out in parallel, the shift register is operating as a *serial-to-parallel converter.* If the data is loaded in parallel and shifted out serially, the shift register is operating as a *parallel-to-serial converter.*

Binary Coding of Numbers in Shift Registers

In the arrangement of Fig. 4-2, four bits can be stored in the register. To accommodate large binary numbers, a method called *weighted coding* is used to represent the FF outputs. Weighted coding has been noted previously; it consists of assigning different values (weights) to each FF. For example, in a 5-bit register, the FF on the right will be assigned a weight of 1. The next FF to the left will be assigned a weight of 2, and the remaining three are weighted 4, 8, and 16, as depicted in Fig. 4-3. When any one of the FFs is set, the stored number is determined by the weight assigned to the FF. For example, if the register readout is 01101, the stored number is 13.

The weighted code exemplified in Fig. 4-3 is commonly called the 8421 *binary code;* one version of this code is termed the *binary coded decimal, or BCD code.* The BCD code employs groups of 4 bits to represent numbers from 0 through 9. Thus, the decimal number 15 is written 1111 in 8421 binary code, but is written 0001 0101 in BCD code. (BCD code is widely used in digital calculators). The 8421 code is the most basic code used in digital systems. Various other codes will also be encountered in specialized digital equipment.

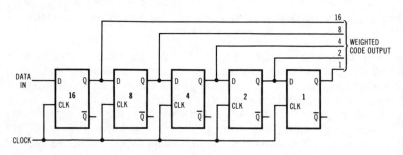

Fig. 4-3. Example of weighted coding of flip-flops in a shift register.
(Courtesy Hewlett-Packard Co.)

Universal Shift Register

Shift registers may be configured for serial-in/serial-out, parallel-in/parallel-out, shift-right, or shift-left operation. A universal shift register, such as shown in Fig. 4-4, is configured to shift data in either direction, to load serially or to load in parallel, and to output data either serially or in parallel. Separate shift-left and shift-right inputs are provided, with right serial output. Mode control terminals permit selection of left-shift or right-shift operation. In this example, an asynchronous clear input is also provided, whereby the contents of the register can be erased independently of the clock.

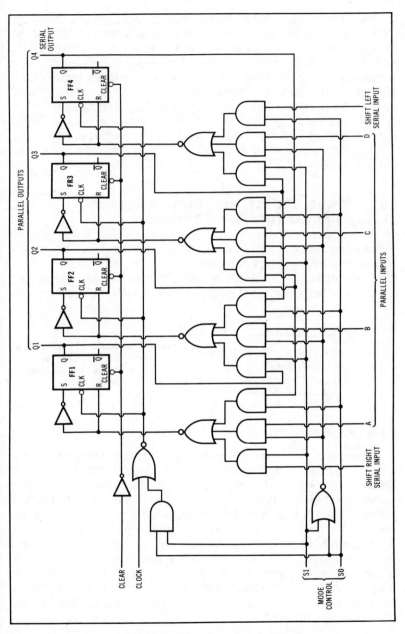

Fig. 4-4. Configuration for a universal 4-bit shift register.
(Courtesy Hewlett-Packard Co.)

This type of register finds extensive application in converting the format of digital data. Thus, if conversion is desired from serial (single-line) data to parallel (multi-line) data, the serial-in/parallel-out mode is employed. The register enters a given number of bits on a single line; in turn, these bits are made available simultaneously on the four output lines. On the other hand, parallel/serial conversion is accomplished by utilizing a parallel-in/serial-out mode of operation. Data on the four parallel input lines is loaded into the register, and then is shifted out from the single output terminal by application of an appropriate number of clock pulses.

Logic Probe, Pulser, and Current Tracer In Register Malfunction Checks

The logic probe, logic pulser, and current tracer are the basic instruments for investigating malfunctions in shift registers. With reference to Fig. 4-5, the logic probe and pulser are applied to determine whether grounds (short circuits to ground) are present.

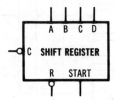

Fig. 4-5. Example of logic probe, pulser, and current tracer application in shift-register checkout.

Case History*

In the example in Fig. 4-5, outputs A, B, C, and D are low; other circuits appear to be normal. First, the probe and pulser were used to make sure that A, B, C, and D aren't grounded (probe and pulse each pin—if ungrounded, the states will be changed by the pulses). Next, other pins on the IC were probed to check for normal/abnormal indications. Then, current was measured at pins A, B, C, and D by pulsing each pin, and tracing to see if current flow is indicated from the pulser to the shift register. In this case, all signals were normal except A, B, C, and D. They were stuck low, suggesting an internal failure in the IC, and not in the circuits connected to it. In this situation, the technician concludes that the IC should be replaced.

---

*Courtesy Hewlett-Packard Co.

Logic Comparator Use in Register Malfunction Checks

With reference to Fig. 4-6, a logic comparator functions by checking the response of a known good IC against the response of a suspect IC of the same type. An example of comparator testing follows.

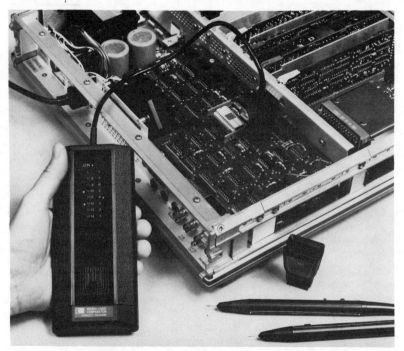

Fig. 4-6. Logic comparator shows discrepancies in logic states between a reference IC and a suspect IC.
(Courtesy Hewlett-Packard Co.)

The fault appears to be pin 13 of A1U8 being held low (see Fig. 4-7). The node consisting of A1U8 pin 13, A2U1 pin 13, and A2U2 pin 4 happens to be spread over two pc boards. The logic comparator was used to find the faulty node. In turn, the comparator identified A1U8 as bad; the troubleshooter then proceeded to check the node with a logic pulser and probe. It was apparent that the node was stuck low. Finally, pulsing and current tracing at A1U8 pin 13 indicated that current was flowing toward pc board A2, and that A2U1 was sinking current and holding the node low.

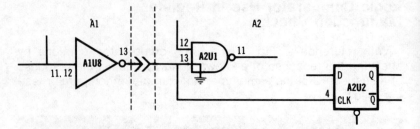

Fig. 4-7. Logic comparator is used to locate a faulty node.
(Courtesy Hewlett-Packard Co.)

Consequently, A2U2 was not being clocked. In summary, the logic comparator immediately located the failure (A1U8 pin 13), but it required the current tracer to track down the fault to A2U1 pin 13.

Section 5

COUNTERS

The operation of a binary counter (Fig. 5-1) is visualized to best advantage by means of a timing diagram, as shown in Fig. 5-2. These are the operating waveforms of a 4-bit binary counter, as would be displayed on the screen of a multichannel oscilloscope. In Fig. 5-2, FF1 is a logic 1 just before FF2 changes state. Logic 1 from FF1 is present only on alternate clock pulses, and FF2 toggles only on alternate clock pulses. Flip-flop FF3 can toggle when both FF1 and FF2 are logic 1. Similarly, FF4 can toggle when FF1, FF2, and FF3 are logic 1.

Down Counter

A down counter starts with a maximum count, such as 1111, and decrements by 1 with each input pulse. Thus, the count sequence is 1111, 1110, 1101, 1100, 1011, 1010, 1001, 1000, 0111, 0110, 0101, 0100, 0011, 0010, 0001, 0000, 1111, and so on. With reference to Fig. 5-3, successive flip-flops are driven by the $\overline{Q}$ output from the preceding flip-flop. (This is the reverse configuration of an up counter, in which successive flip-flops are driven by the Q output from the preceding flip-flop.) Note that in order to preset the counter to 1111, all of the direct set inputs are tied together. Thereby, the counter can be preset at any time. Note that after the count reaches 0000, the next input pulse causes it to start the cycle over again and read 1111. This return could be called "down overflow" as shown in Fig. 5-4.

Up Down Counter

With reference to Fig. 5-5, an AND-OR gate with an inverter can be added to a counter configuration so that it will either count up

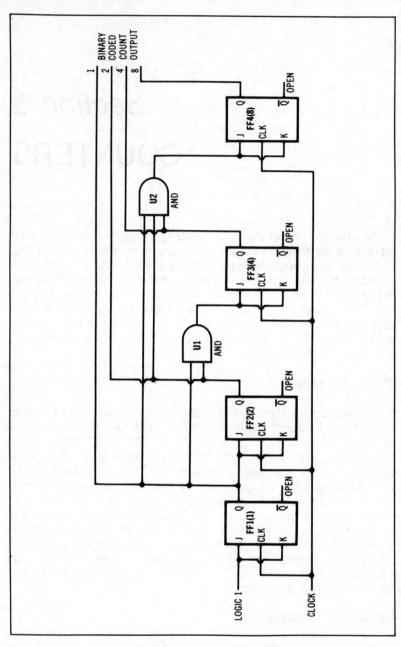

Fig. 5-1. Typical 4-bit binary counter.

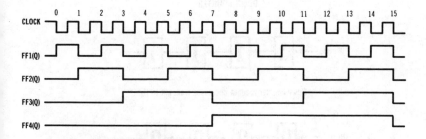

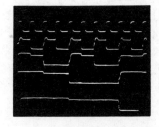

Fig. 5-2. Timing diagram for a 4-bit binary counter.
(Courtesy Fairchild Camera and Instrument Corp.)

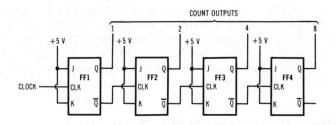

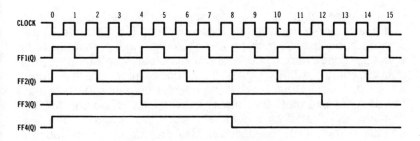

Fig. 5-3. Basic down asynchronous counter configuration and timing diagram. *(Courtesy Hewlett-Packard Co.)*

DOWN COUNTER

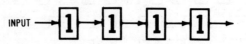

Down-counter readout decreases from left to right.

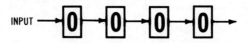

Down-counter overflow causes 1111 read out.
Count then decreases from left to right.

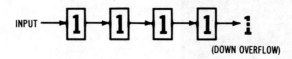

Fig. 5-4. Visualization of "down overflow."

or count down. If the count mode input is logic-high, gate B is open, so that an up count is obtained. On the other hand, if the count mode input is logic-low, gate A is open, so that a down count is obtained. As noted above, all of the direct set inputs may be tied together in order to preset the counter when the count mode input is logic-low. Similarly, all of the direct reset inputs may be tied together in order to preset the counter when the count mode input is logic-high.

Up Down Counter in Digital Color TV

An example of an up down counter and decoder application in a digital color-tv receiver is shown in Fig. 5-6. Decoders, described subsequently, use an arrangement of gates that converts a binary coded input into a specific output. For example, if a counter outputs a 0001 signal, the decoder interprets this signal as a 1 decimal output. Next, if the counter outputs a 0010 signal, the decoder interprets this signal as a decimal 2 output. A 0011 signal is interpreted as a decimal 3 output, and so on. With four counter output lines, the decoder may provide up to 16 decimal output lines. In the example of Fig. 5-6, these decimal output lines correspond to tuner channels.

106

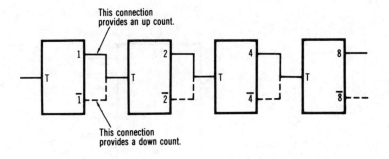

This connection provides an up count.

This connection provides a down count.

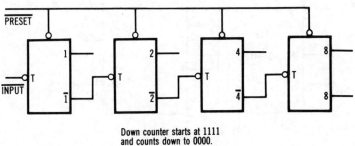

Down counter starts at 1111 and counts down to 0000.

"Down overflow" then returns the count to 1111.

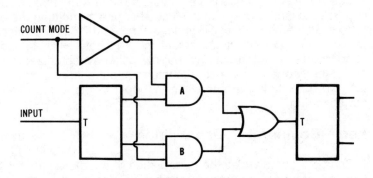

Basic up down counter configuration.

Fig. 5-5. Development of an up down counter.

Strobe Pulse

A strobe pulse serves to enable a gate, so that data present on an input line will not pass until the proper time. For example, a

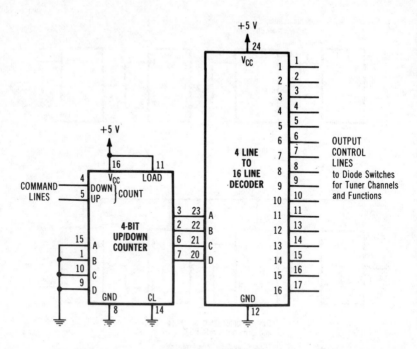

Fig. 5-6. Up down counter application in digital color tv receiver.

strobe pulse is often used to eliminate a race problem; as seen in Fig. 5-7, the strobe pulse enables an AND gate, permitting the data to pass after a short delay. This delay time permits all of the inputs to attain their correct states before the AND gate will respond. A strobe pulse is a comparative narrow pulse synchronized to the clock; the strobe pulse is said to interrogate the input condition of the AND gate during a time when the input levels to the gate are not changing.

Logic Comparator Capabilities

Chart 5-1 summarizes the capabilities of a logic comparator in testing gate combinations (combinatorial logic), flip-flops, shift registers, and related devices. A logic comparator provides a quick and easy checkout of a counter whenever a reference IC is available. However, the comparator must often be supplemented by a logic pulser and probe, a current tracer, or an oscilloscope. A comparator will indicate which pin(s) may respond with incorrect logic states, but it cannot show whether the IC is defective or the associated circuitry is defective.

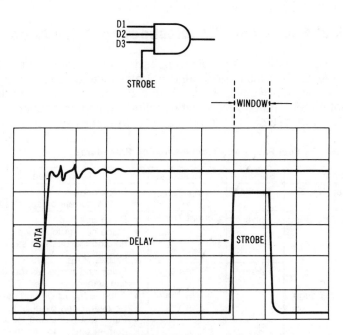

Fig. 5-7. Example of strobe pulse.

**Chart 5-1. Comparator Capabilities Chart
(16 or Less Pin Dual-in-Line DTL, TTL)**

Combinatorial Logic (AND, NAND, XOR, etc.)	Excellent. This also includes expandable ICs. This category includes the vast majority of ICs in use.
Sequential Logic (Flip-Flops)	Excellent. Reference and test IC should be synchronized by a pulse on the "Reset" input.
Memories, Shift Registers	Excellent. Clip must be attached long enough for reference and test IC to contain the same information.
Low Power TTL	Good. The Comparator is an input load of 5 for the driving device.
One-shots	Usually good. Since reference and test IC share the RC timing components, circuit timing can be affected.
Open collector and Tri-state Logic	Usually poor. When outputs are bussed together a good gate is constrained to operate improperly and this will be indicated by the Comparator.
Expanders, Analog and Linear ICs	No. Their outputs are analog and cannot be tested by the Comparator.
MOS Devices	No. They require different power supplies exceeding the 7V input limit and will damage the Comparator.

Courtesy Hewlett-Packard Co.

Typical Faults and Troubleshooting Procedures

Table 5-1 summarizes typical faults in digital equipment, with the preferred test instruments and troubleshooting procedures. In some situations, results obtained from use of the pulser, probe, current tracer, and clip lead to the malfunctioning area, but do not pinpoint the defect. In such a case, a triggered-sweep oscillo-

Table 5-1. Typical Faults and Troubleshooting Procedures

Fault	Stimulus	Response	Test Method
Shorted Node[1]	Pulser[2]	Current Tracer	• Pulse node • Follow current pulses to short
Stuck Data Bus	Pulser[2]	Current Tracer	• Pulse bus line • Trace current to device holding the bus in a stuck condition
Signal Line Short to V_{cc} or Ground	Pulser	Probe Current Tracer	• Pulse and probe test point simultaneously • Short to V_{cc} or Ground cannot be overridden by pulsing • Pulse test point, and follow current pulses to the short with tracer
V_{cc} to Ground Short	Pulser	Current Tracer	• Remove power from test circuit • Disconnect electrolytic bypass capacitors • Pulse across V_{cc} and ground using accessory connectors provided • Trace current to fault
Suspected Internally Open IC	Pulser[2]	Probe	• Pulse device input • Probe output for response
Solder Bridge	Pulser[2]	Current Tracer	• Pulse suspect line(s) • Trace current pulses to the fault (Light goes out when solder bridge passed)
Sequential Logic Fault in Counter or Shift Register	Pulser	Clip	• Circuit clock de-activated • Use Pulser to enter desired number of pulses • Clip onto counter or shift register and verify device's truth table

[1]A node is an interconnection between two or more ICs.
[2]Use the Pulser to provide stimulus, or use normal circuit signals, whichever is most convenient.

Courtesy Hewlett-Packard Co.

scope is used to close in on the defect. A logic clip is placed over an IC as illustrated in Fig. 5-8. In Fig. 5-9, shift register inputs are pulsed with the logic pulser; the logic clip shows resulting high and low states of the IC terminals.

Fig. 5-8. Logic clip is placed on the IC under test.
(Courtesy Hewlett-Packard Co.)

Fig. 5-9. Troubleshooting with the logic pulser and logic clip.
(Courtesy Hewlett-Packard Co.)

Section 6
DECODERS AND ENCODERS

A decoder is a device that converts coded data into another form, such as a binary-to-decimal decoder. The input data is generally loaded in parallel; in turn, a single corresponding output is obtained, as depicted in Fig. 6-1. Another example of a widely used decoder is the BCD-to-7 segment configuration which provides a specified combination of outputs for each BCD input; it is used to drive a 7-segment readout device (see Fig. 6-2).

An encoder is a logic configuration that produces coded com-

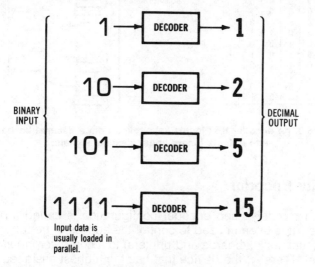

Fig. 6-1. Typical decoder operation.

| 0 | 1 | 2 | 3 | 4 | 5 | 6 | 7 | 8 | 9 | 10 | 11 | 12 | 13 | 14 | 15 |

Fig. 6-2. Numerical designations and resultant displays for a 7-segment readout device.

binations of outputs from discrete inputs. As an illustration, Fig. 6-3 shows an arrangement for changing decimal keystrokes into a binary coded decimal (BCD) output. In BCD code, the decimal digit 7 corresponds to 0111. Therefore, the 7 keystroke applies logic-high levels to the 4, 2, and 1 OR gates. Note that only one decimal digit at a time may be applied to the encoder input; only one BCD output is logic-high at any given time. This type of encoder is usually built into the microcomputer keyboard.

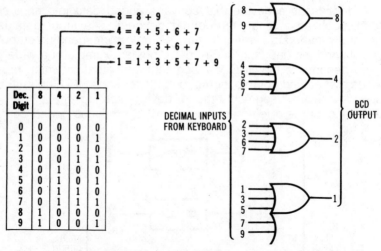

8 = 8 + 9
4 = 4 + 5 + 6 + 7
2 = 2 + 3 + 6 + 7
1 = 1 + 3 + 5 + 7 + 9

Dec. Digit	8	4	2	1
0	0	0	0	0
1	0	0	0	1
2	0	0	1	0
3	0	0	1	1
4	0	1	0	0
5	0	1	0	1
6	0	1	1	0
7	0	1	1	1
8	1	0	0	0
9	1	0	0	1

DECIMAL INPUTS FROM KEYBOARD

BCD OUTPUT

Fig. 6-3. An encoder that inputs keystrokes from a decimal keyboard and outputs binary coded decimal signals.

Priority Encoder

Another widely used encoder configuration is called a *priority encoder*. It is often utilized to control the access of peripheral devices (such as keyboards and printers) to a computer input/output channel. Thereby, the device that has the highest assigned priority gains access to the channel before any device with a lower

114

assigned priority. A typical priority encoder may have eight input lines and three output lines, as shown in Fig. 6-4. Whenever one input line, such as $\overline{6}$ (6 active-low) is enabled by the peripheral device connected to it, the priority encoder outputs a binary 6 count (110) to the computer. When two or more input lines, such as $\overline{3}$, $\overline{6}$, and $\overline{7}$, are enabled, the priority encoder outputs a binary count corresponding to the highest order line—in this example, $\overline{7}$. If only one input line is activated at a time, the priority encoder responds as an 8-line-to-binary encoder.

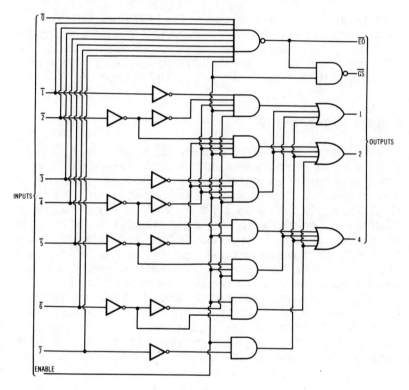

Fig. 6-4. Priority encoder configuration. *(Courtesy Hewlett-Packard Co.)*

MSB and LSB

Two important terms that the troubleshooter should be familiar with are the *most significant bit* (MSB) and *least significant bit* (LSB). Binary weights were noted previously; a logic 1 or 0 stored in an FF with the smallest weight (FF marked 1) is the least signifi-

115

cant bit. On the other hand, the bit stored in an FF with the highest weight is the most significant bit. MSB and LSB are also used to identify the bits in any bit stream. As an illustration, the bit stream for the decimal number 5 corresponds to the binary number:

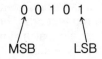

There are five bits in this stream. The MSB does not enter into the absolute value of the binary number 101. In some circuitry, the MSB would be "meaningless." However, in other circuitry, the MSB might be a *sign bit,* denoting that 101 is positive; in this type of circuitry, 100101 would denote that 101 is negative.

Digital Clock Signals

Clock signals have been previously described. Any digital system that employs flip-flops includes a clock. Generally, the clock signal is a square wave. All flip-flops, shift registers, counters, and other timed devices in the system operate from the same clock. Hence, this is called a *master clock* or a *system clock.* It synchronizes the digital system, and it also determines how fast the system operates. Every timed device has a rated maximum clock rate or frequency. For example, a shift register might be rated for a clock frequency up to 5 MHz. In theory, a clock waveform is a true square wave. However, in practice, it deviates from the ideal—sometimes slightly, and at other times substantially. Overshoot and ringing are a common form of clock waveform distortion; slow rise time is another form of distortion. Excessive distortion can lead to system malfunction. A high-performance oscilloscope is used to check clock waveforms.

Section 7
MULTIPLEXERS

A multiplexer is a configuration wherein the data on multiple input lines is processed and rearranged for outputting on a single line. In general, multiplexing processes the mixing of signals from multiple sources into a lesser number of outputs. A multiplexer can also be operated to select one input and to continuously gate the data from the particular input through to the output. When utilized in this manner, the arrangement is called a *data selector*.

With reference to Fig. 7-1, an eight-channel multiplexer is shown in which the data rearrangement or multiplexing is controllable by means of three input signals. In other words, the logic state present on any selected input, D0–D7, is placed in a chosen sequence on the output line T. Accordingly, the address inputs A0, A1, A2 are stepped in binary code to select a particular input at the proper time. As an illustration, an address input A0=0, A1=0, A2=0 will place the logic state that is present on D0 into

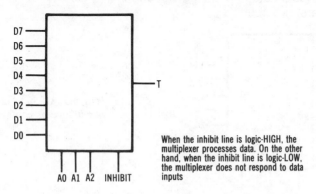

When the inhibit line is logic-HIGH, the multiplexer processes data. On the other hand, when the inhibit line is logic-LOW, the multiplexer does not respond to data inputs

Fig. 7-1. An eight channel multiplexer configuration.

the output line T. Also, an address input A0=1, A1=0, A2=0 would place the logic state that is present on D1 into the output line T.

In theory, a multiplexer may mix signals from multiple sources in various ways into a smaller number of outputs. Thus, although the foregoing example of mixing eight data lines into a particular sequence on a single output line merely represents one widely used mode. Other multiplexers encountered in digital logic diagrams serve to multiplex 4 lines into 1 output line, 16 lines into 1 output line, or digital word multiplexers that operate to multiplex three 4-bit wide parallel words into a single 4-bit wide output. With reference to Fig. 7-1, a fixed address may be applied, whereby the data from one input line is continuously channeled through to the output line T. This operating mode of a multiplexer is termed a *data selector*.

Demultiplexers

A demultiplexer has the opposite function of a multiplexer. That is, a demultiplexer receives data from a single source and distributes the data in accordance with a particular pattern into several output lines. For example, the demultiplexer shown in Fig. 7-2 has eight output lines and can accomplish the reverse data processing provided by the multiplexer depicted in Fig. 7-1. In other words, serial data enters via input line T, the binary address on A0, A1, and A2 is sequentially stepped from 000 through 111, and in turn the inputted data appears sequentially on output lines D0 through D7. Commercial demultiplexer IC packages are available to distribute data from a single line into 2, 4, 16, and various other numbers of outputs.

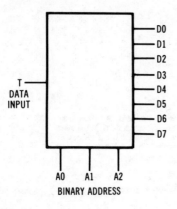

Fig. 7-2. An eight channel demultiplexer configuration.

Note that instead of being demultiplexed, data on the single input line (Fig. 7-2) can be gated by means of the A0, A1, and A2 address lines through any one of the output lines continuously. Then, at a later time, the address may be changed to continuously gate the data through another output line. The demultiplexer is called a *data distributor* in this mode of operation. Observe also that a demultiplexer can be used as a decoder. In other words, if the input line T (Fig. 7-2) is maintained logic-high, the outputs will always represent the binary count corresponding to the logic levels on A0, A1, and A2. For example, logic levels 101 result in output line D5 going high. (Line T becomes the enable function.) It is evident that a demultiplexer with 4 address inputs and 16 outputs can be operated as a 4-line to 16-line decoder.

Analytical Versus "Shotgun" Approach

When the troubleshooter is confronted by a puzzling malfunction in a digital system, much time and effort can often be saved by taking an analytical approach. Otherwise, ICs must be checked at random; elaborate digital systems may contain dozens of ICs. Even after all ICs have been verified as workable in a go/no-go test procedure, the system may still refuse to operate normally. In turn, an analytical approach is the only effective method of contending with a "tough dog." For example, sharp noise pulses or other narrow "glitches" often cause malfunction of digital circuitry. A glitch may be caused by radiated noise, power-line coupled noise, "sliver" pulses due to abnormal timing-skew input signals to gates, or similar faulty circuit actions. Zeroing in on a glitch can be a demanding job, even when the troubleshooter is certain that the glitch is present. Analysis of glitch trouble symptoms requires a *thorough understanding* of the *digital circuitry*, so that the troubleshooter can reason his or her way to the trouble area. In turn, appropriate test signals can be applied to narrow down the fault location, and possibly to pinpoint the source. Adequate test equipment is also necessary—in difficult situations, the difference between success and failure can be the "edge" that a professional oscilloscope provides over an economy-type oscilloscope.

Preliminary Troubleshooting Steps

Documentation of digital equipment is as essential as adequate test instruments in puzzling trouble situations. The troubleshooter must study the block diagrams and the logic diagrams of schematics. A well-documented system is provided with a system block diagram, and notes on the operation of subsystems. Test inputs may be provided. Functions of address lines and data buses (if utilized) are generally summarized. Waveform charts are frequently included, with notes on signal processing. Trouble-shooting charts may also be provided. The troubleshooter should determine as many symptoms of malfunction as possible, so that the basic problem can be fully defined. Good circuit boards should be substituted for suspected defective circuit boards, when feasible. A "finger test" will sometimes identify an over-heated failed component. Signal tracing should start in large steps until a normal waveform is found; then, the signal tracing may proceed in small steps. It is often helpful to select points for probing which are outputs of simple logic gates. Probe device pins, not socket pins.

Intermittent Trouble Symptoms

Digital logic troubleshooting proceeds under three general approaches: signal substitution, signal tracing, and aggravation. Aggravation techniques are used in troubleshooting intermittents to locate the area of malfunction. Without the use of excessive force, the technician may twist and pull connections, cables, plugs, and plug-in units while monitoring the output of the equipment with a logic probe. It is sometimes helpful to "wipe" the handle of a plastic screwdriver across the back of a suspected row of modules to initiate the intermittent. Individual modules may be tapped, to narrow down the possibilities; suspected modules can be wiggled up and down. Although most intermittents are mechanical, some are thermal. A thermal intermittent can often be initiated by blowing hot air into suspected areas, followed by a stream of cold air.

Section 8
COMPARATORS

Digital comparators are used in various applications; for example, a comparator functions to compare the values of two binary numbers, A and B, and to determine whether one is larger than the other, or if both numbers are equal. Comparators are employed in various adders and subtracters; they are also utilized in digital control circuitry, wherein the control function that is generated depends upon the comparative values of two or more inputs. Binary addresses are compared for relative magnitude in some applications. An example of a commercial 4-bit quad exclusive-NOR comparator is shown in Fig. 8-1. This comparatively simple configuration determines only equality or inequality. If both inputs are 1, or if both inputs are 0, the output is 1. On the other hand, if one input is 1 and the other input is 0, the output is zero.

Another commercial example of a more elaborate comparator is also depicted in Fig. 8-2. This is a 5-bit configuration; it determines whether one digital word is equal to, greater than, or less than another digital word. The comparator responds to word inputs as specified in the truth table. An active-low enable input is included, whereby the comparator can be turned "on" or "off." In this example, the A digital word is 10011, and the B digital word is 11010; the enable input is 0, with the result that the A<B output is 1, the A>B output is 0, and the A=B output is 0.

Polarity Comparator

A polarity (sign) comparator indicates whether two input voltages have the same polarity, or whether they have opposite polarities. Two flip-flops and two AND-OR gates are utilized in the configuration, as shown in Fig. 8-3. If V1 and V2 are both positive,

121

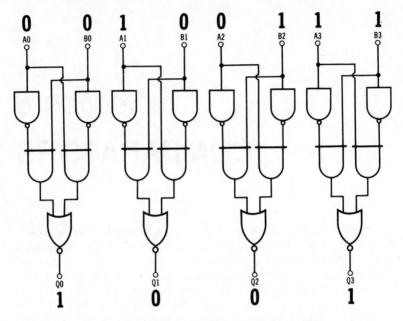

Fig. 8-1. A 4-bit quad exclusive-NOR comparator.

gate A will have both inputs positive, and the "like polarities" output will go logic-high. Or, if V1 and V2 are both negative, gate B will have both inputs positive, and the like polarities output will go logic-high. On the other hand, if V1 is positive and V2 is negative, gate C will have both inputs positive, and the unlike polarities output will go logic-high. Also, if V1 is negative and V2 is positive, gate D will have both inputs positive, and the unlike polarities output will go logic-high.

Note that when gate A has both inputs positive, gate B has both inputs negative, gate C has a positive input and a negative input, and gate D has a positive input and a negative input. Accordingly, only gate A produces a logic-high output. Also, when gate A has both inputs negative, gate B has both inputs positive, gate C has a positive input and a negative input, and gate D has a positive input and a negative input. In turn, only gate B produces a logic-high output. The same general principle applies when V1 and V2 have opposite polarities—only one gate can go logic-high for a particular combination of input polarities.

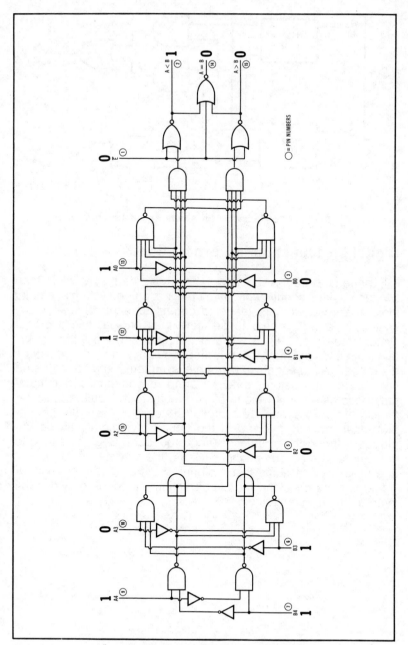

Fig. 8-2. A 5-bit comparator configuration.

123

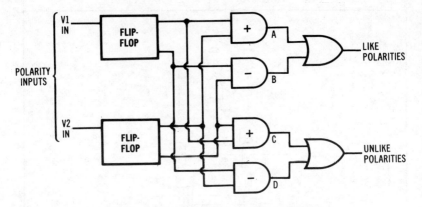

Fig. 8-3. Configuration for a polarity comparator.

Digital Troubleshooting Precautions

Experienced troubleshooters generally agree that no more than two technicians should work on a malfunctioning system at a time; more than two workers leads to confusing situations. In many cases, there will appear to be more than one fault in a digital system (see Fig. 8-4). It is good practice to correct the "easy" faults first; sometimes the puzzling trouble symptoms will then disappear. All recognizable trouble symptoms should be listed and "sized up" before work is started; a common denominator may become evident which points directly to the fault. It is sometimes helpful to subdivide a digital system; for example, peripherals that are not involved in the trouble symptom may be unplugged during the troubleshooting procedure. Beginners tend to make unjustified assumptions, such as "the defect must be in . . ." It is illogical to assume that a signal exists or that it is a *normal signal* without checking the circuit action—a scope provides the most detailed information.

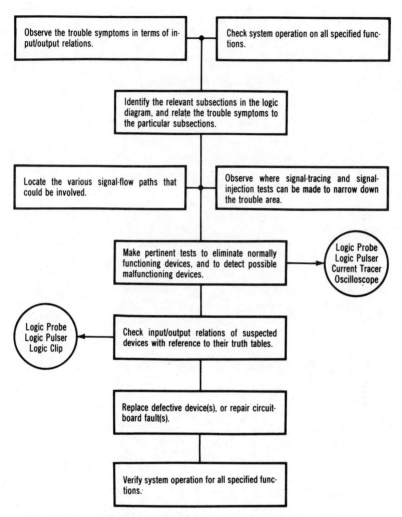

Fig. 8-4. Use of digital logic diagrams in troubleshooting procedures.

125

Section 9
PARITY GENERATOR/ CHECKERS

Since digital systems are not completely reliable, a basic requirement concerns recognition of processing errors in digital data. For example, in a system that uses a tape recorder for mass storage of data, it is essential to have an indicator for incorrect transfer of a binary word to or from the recorder. For example, if a particular bit is incorrectly read as a 0 instead of a 1 in a digital word, an erroneous answer would be obtained. In turn, a suitable method of checking and detecting an erroneous bit is required.

A widely used method of error detection in digital words is to provide counting of the number of logic-1 bits in a 7-bit character, as follows:

Character or Command	Code
A	1000001
B	1000010
Z	1011011

An even number of logic-1 bits is present in the digital code word for A and for B; an odd number of logic-1 bits is present in the digital code word for Z. When the number of logic-1 bits is odd, one more logic-1 bit is added to the group to form an 8-bit word that contains an even number of logic-1 bits. On the other hand, when the number of logic-1 bits in the 7-bit character is even, a logic-0 bit is added in the eighth position, so that the 8-bit digital code word contains an even number of logic-1 bits as follows:

Character or Command	Digital Code Word
A	01000001
B	01000010
Z	11011011

↑
Parity
Bit

Accordingly, every 8-bit digital word contains an even number of logic-1 bits. In turn, when these 8-bit words are transferred into or out of a memory, a receiving device counts the logic 1s in each word to verify that the number is even. In case of a malfunction, an odd number of logic-1 bits in a word causes an error signal to be triggered. This process is called *parity checking* and the particular method that has been described is termed *even parity*. Note that odd parity checking is also employed in digital systems—in this case, each valid binary word normally contains an odd number of logic-1 bits.

Although a system malfunction might cause two logic-1 bits in the same 8-bit digital word to drop out, this is not very probable from a statistical viewpoint. Of course, if a double drop-out should occur in the same binary word, the parity check would pass the word as valid. Note that by adding more bits to each binary word, and by elaborating the encoding and detection network, error checking processes can be made more reliable. A sophisticated arrangement can detect which bit in a binary word is erroneous. In addition, the system can be designed so that the erroneous bit is automatically corrected; these functions are commonly provided in large computing systems.

Parity generator/checkers are widely utilized in data bus arrangements, as depicted in Fig. 9-1. A *bus* is simply one, or many, conductors (such as pc conductors) used as a path over which digital information is transmitted from any of several sources to any of several destinations. A *data bus* carries digital data to or from a number of different locations. The term *data* denotes basic elements of information which can be processed or produced by a digital computer. A fundamental configuration employing an odd-parity generator is shown in Fig. 9-2.

Although the transmitting section is programmed to output 0000,

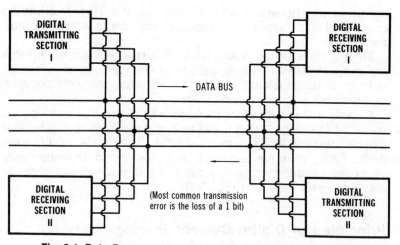

Fig. 9-1. Data flow may proceed in either direction along a bus.

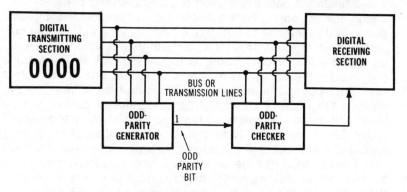

COMPARATIVE EVEN — AND ODD — PARITY BITS			
DECIMAL	BCD	EVEN-PARITY BIT	ODD-PARITY BIT
0	0000	0	1
1	0001	1	0
2	0010	1	0
3	0011	0	1
4	0100	1	0
5	0101	0	1
6	0110	0	1
7	0111	1	0
8	1000	1	0
9	1001	0	1

Fig. 9-2. A 4-bit digital data transmitting system that includes an odd-parity generator at the transmitting terminal and an odd-parity checker at the receiving terminal.

the odd-parity generator will normally output a 1 parity bit. If the checker should output a 0, a lack of odd parity would be indicated.

Short circuits occasionally occur in buses. For example, there are many troubleshooting situations wherein a faulty circuit node is found to be "stuck at," and if there are many elements common to this node, a tough-dog problem can result.

A parity tree is named for its resemblance to a tree with various branches (see Fig. 9-3). It is defined as a group of XOR gates that can be used to check a number of input bits for either odd or even parity. Parity trees are used both to check and to generate parity wherever a redundant bit is added to a digital word in order to check for error (see also Fig. 9-4).

Principles of Digital Current Tracing

It is helpful to note the amount of current flowing between logic circuit devices during high and low states when trouble symptoms occur. With reference to Fig. 9-5, I_{OH} denotes the high output current (output at logic 1); in TTL circuits this is about 40 μA. I_{OL} denotes the low output current (output at logic 0); in TTL circuits this is normally about 1.6 mA for a fan-out of one. If the output becomes shorted, the current may be as high as 55 mA. Thus, a short at the input of U2 would cause 55 mA of current to flow toward U2 when U1 tries to go high; no logic state change would result due to the short. Note that when the output of U1 is at a logic-low state, Q3 is a saturated transistor and draws about 1.6 mA. Transistor Q3 has the capability to drive up to 10 gates, and accordingly to sink up to 16 mA of current in normal operation. When U1 goes logic-low its output transistor sinks 1.6 mA of current from each inverter. Therefore, the normal current is about 8 mA. A short anywhere along the node comprising the U1 output and the inputs to inverters U2–U6 will greatly alter both the magnitude and the direction of current flow (see Fig. 9-6). Basic troubleshooting rules are: (1) In a known bad node, the current usually exceeds the other currents on the pc board by a wide margin. (2) Determine the *source* and the *sink* of current flow; in turn, the faulty component can often be quickly pinpointed with the logic pulser.

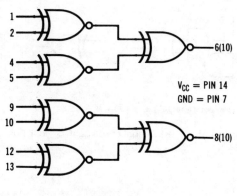

$6 = 1 \oplus 2 \oplus 4 \oplus 5$
where $X \oplus Y = \overline{X} \cdot \overline{Y} + X \cdot Y$

(A) Dual 4-bit parity tree.

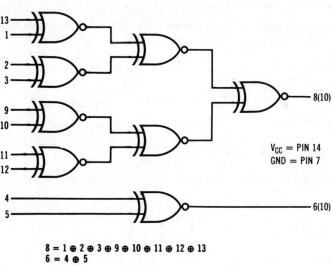

$8 = 1 \oplus 2 \oplus 3 \oplus 9 \oplus 10 \oplus 11 \oplus 12 \oplus 13$
$6 = 4 \oplus 5$

where $X \oplus Y = \overline{X} \cdot \overline{Y} + X \cdot Y$

(B) 8-bit parity tree.

Fig. 9-3. Examples of parity trees.

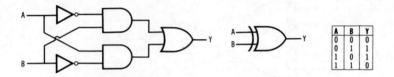

Logic equation of two-bit XOR gate: $Y = \overline{A}B + A\overline{B} = A \oplus B$

(A) Logic diagram of XOR gate. (B) Symbol of XOR gate. (C) Truth table for 2-bit XOR gate.

A	B	Y
0	0	0
0	1	1
1	0	1
1	1	0

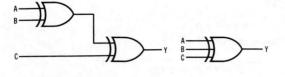

Logic equation of three-bit XOR gate: $Y = A \oplus B \oplus C$

(D) Logic diagram of 3-bit XOR gate. (E) Symbol of 3-bit XOR gate. (F) Truth table for 3-bit XOR gate.

A	B	C	Y
0	0	0	0
1	0	0	1
1	1	0	0
1	1	1	1
0	0	1	1
0	1	1	0
1	0	1	0
0	1	0	1

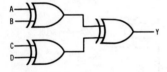

Logic equation of four-bit XOR gate: $Y = A \oplus B \oplus C \oplus D$

(G) Logic diagram of 4-bit XOR gate.

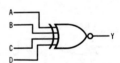

(H) Symbol of 4-bit XOR gate.

A	B	C	D	Y
0	0	0	0	0
0	0	0	1	1
0	0	2	0	2
0	0	1	1	0
0	1	0	0	1
0	1	0	1	0
0	1	1	0	0
0	1	1	1	1
1	0	0	0	1
1	0	0	1	0
1	0	1	0	0
1	0	1	1	1
1	1	0	0	0
1	1	0	1	1
1	1	1	0	1
1	1	1	1	0

(I) Truth table for 4-bit XOR gate.

Fig. 9-4. Summary of XOR gate circuitry.

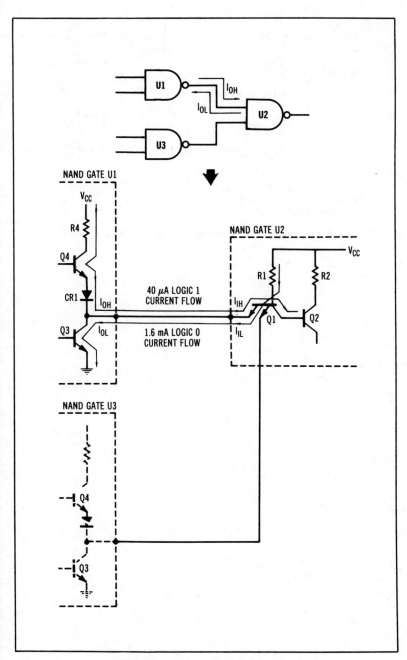

Fig. 9-5. I_{OL} and I_{OH} current relations.

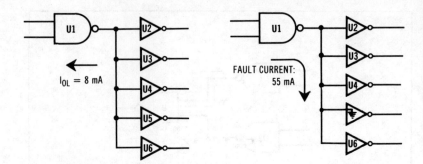

(A) A short-circuited node greatly changes the magnitude and direction of current flow.

(B) Example of shorted input to U5.

Fig. 9-6. Basic digital current tracing. *(Courtesy Hewlett-Packard Co.)*

Section 10
SCHMITT TRIGGERS AND MONOSTABLE MULTIVIBRATORS

A Schmitt trigger is a bistable multivibrator in which the output state changes with great rapidity when the input voltage exceeds an upper voltage value or falls below a lower voltage value. The difference between these two voltages is called the hysteresis voltage; in a typical circuit, the hysteresis voltage is 800 mV. A Schmitt trigger is often described as a waveform squaring circuit; regardless of the input waveshape, the output waveform from a Schmitt trigger is always a rectangular wave.

Monostable Multivibrators

A monostable multivibrator, also called a one-shot multivibrator, or start-stop multivibrator, is a rectangular-wave generator that occupies a functional position intermediate to the free-running multivibrator and the bistable multivibrator. A monostable multivibrator has only one stable state. It can be triggered to change state, but only for a predetermined interval, after which it returns to its resting state. A monostable multivibrator is sometimes called a pulse regenerator. It produces output pulses that are independent of the input trigger repetition rate. When an input trigger is applied, a monostable multivibrator flips to its unstable state for a period which is determined by an RC time constant. Then the monostable multivibrator returns to its original stable state. In the example of Fig. 10-1, a complementary pair of narrow output

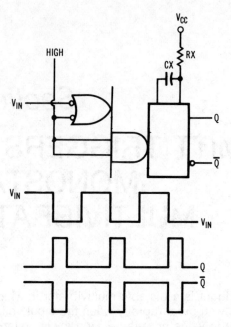

Fig. 10-1. Operation of a monostable multivibrator.

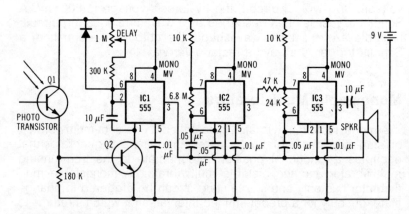

Fig. 10-2. Monostable multivibrator application in a delayed light alarm.

pulses is produced on each trailing edge of the input signal. The width of the output pulse depends on the values of Cx and Rx.

Monostable Multivibrator Applications

Elaborate types of digital-logic probes include a monostable multivibrator for operation as a pulse stretcher. In other words, glitches and other very narrow pulses may be encountered in digital systems. Because a very narrow pulse has much less energy than a normal pulse, it does not produce a visible indication when a simple logic probe is used. On the other hand, if a multivibrator is included in the probe circuitry, a clearly visible indication will be obtained when a very narrow pulse is inputted. That is, the monostable multivibrator can be triggered by a glitch, and will in turn output a pulse of suitable width for energizing an LED.

Another common application for a monostable multivibrator is shown in Fig. 10-2. This is a home-appliance application, and the unit is basically intended as a refrigerator monitor, to indicate whether the door has been completely closed. When light strikes the phototransistor, there is no immediate response; however, after several seconds, if the light continues to strike Q1, the alarm will start to sound.

Section 11

EXPANDABLE GATES AND DIGITAL LOGIC DIAGRAMS

An expandable gate is similar to a conventional gate except that it includes provision for increasing the number of inputs by addition of a logic block. With reference to Fig. 11-1, the X input is a connection to the output section of the expandable gate. Accordingly, if the X input is pulsed, an output pulse results in the same manner as if AB, CD, EF, or GHI were pulsed. Up to four external gates are typically channeled into the X input.

Analyzing Digital Logic Diagrams

Analysis of a digital logic diagram starts with determination of the system function, or a sectional function. Thus, it may be observed that the section of interest functions as a counter, or a register, or a decoder. If the digital troubleshooter understands the theory of operation for the particular network, he or she can "size up" the diagram with respect to the trouble symptoms and reject various inapplicable assumptions. In turn, the technician can concentrate on the valid approaches and save considerable time. Sometimes there are seemingly several trouble symptoms that actually relate to a single fault. An understanding of the digital circuit action enables the technician to see the interrelations of various trouble symptoms (see Chart 11-1).

As a practical note, "tough dog" trouble symptoms can be caused by very narrow glitches that are invisible on service-type

139

oscilloscopes. This type of glitch is typically caused by a marginal device that has not yet failed catastrophically. A high-performance lab-type oscilloscope, preferably of the storage type, is required to track down a glitch with a very small width. The chief requirements for a visible display are high writing speed (for glitch expansion), and ample beam intensity. If the logic diagram is not too complex, it may be possible for an experienced technician to "size up" the circuit actions and make an "educated guess" concerning the possibility of narrow glitch generation in various subsections.

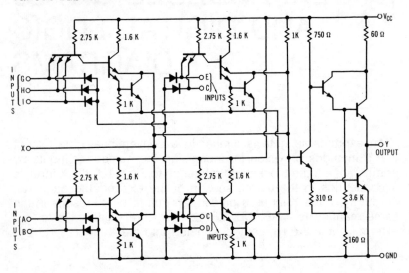

Fig. 11-1. Expandable 2-2-2-3 input AND-OR gate.

Chart 11-1. "Sizing Up" Digital Logic Diagrams

Preliminary Considerations:
1. What is the basic function of the digital system?
2. List the functional subsections.
3. Is the malfunction solid (permanent), or is it intermittent?
4. Have you checked the power-supply voltage, and the ripple?

Evaluation of Trouble Symptoms:
1. Note precisely what happens when the malfunction occurs.
2. How and when does the fault become observable?
3. Is the failure recurrent?
4. Do trouble symptoms appear in a significant sequence?
5. Is there a significant relationship among the trouble symptoms?

Progressive Evaluation:
1. Is there an instruction manual for the digital system?
2. Do you have a maintenance manual?
3. Is a service index with troubleshooting charts available?
4. Do you have a file of case histories?

Planning a Preliminary Approach:
1. Can you work up a system down flow diagram?
2. Note all of the functional subsections that could possibly be involved.
3. Consider signal-injection and signal-tracing tests that can be made to narrow down the possible trouble area.
4. Have all obvious possibilities been checked, such as loose connections, low line voltage, and "cockpit error?"
5. Is a similar digital system available in normal working condition for making comparative tests?

Section 12
INVERTERS, BUFFERS, AND DRIVERS

As noted previously, a buffer is basically an amplifier, and an inverter is bascially an amplifier that reverses the polarity of the input signal. The technician will encounter buffers connected in parallel, and inverters connected in parallel; this is done to increase the signal-power capability of the devices. Some buffers or inverters serve essentially as interfaces; thus, the device may interface a TTL section to a MOS section, or interface a MOS section to a TTL section, as shown in Figs. 12-1 and 2. Other buffers or inverters serve as high-voltage high-current drivers.

A differential line driver converts a single-ended input signal into a double-ended (push-pull) output signal; a double-ended signal is also called a balanced signal (see Fig. 12-3). The advan-

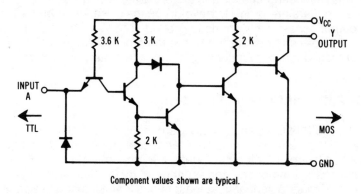

Component values shown are typical.

Fig. 12-1. Logic diagram for an inverter buffer/driver interface.
(Courtesy Fairchild Camera and Instrument Corp.)

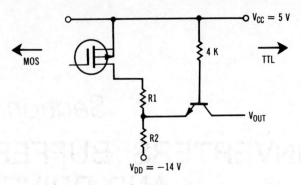

Fig. 12-2. A MOS-to-TTL buffer.

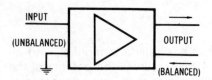

Fig. 12-3. Principle of a differential line driver.

tage of differential drive is minimization of interference due to common mode line noise. Note that a differential driver can also be operated in the single-ended mode by grounding one of its output terminals. Tri-state buffers (Fig. 12-4) are similar to basic buffers, except that they are provided with an enable input. When the enable input is *high,* the buffer operates conventionally. On the other hand, if the enable input is *low,* the buffer is effectively disconnected from the data bus.

Troubleshooting tips for bus drivers are listed in Table 12-1. Like gates, buffers are prone to develop opens or shorts; a buffer input or output occasionally shorts to ground or to V_{cc}. The logic pulser, logic probe, and current tracer are the most useful instruments for troubleshooting conventional buffers or tri-state bus drivers.

RS-232-C Interface Function

The RS-232-C is an EIA standard defining the interfacing between data terminal equipment and data communication equipment utilizing serial binary data interchange. *Mark* and *space* signals are employed, corresponding to familiar *high* and *low* levels. A mark potential is at least minus 3 volts; a space potential is at least plus 3 volts. An RS-232-C line driver converts TTL logic levels to EIA levels for transmission between data terminal equip-

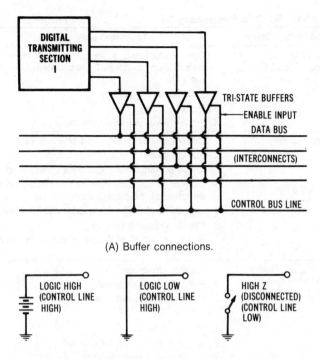

(A) Buffer connections.

LOGIC HIGH
(CONTROL LINE
HIGH)

LOGIC LOW
(CONTROL LINE
HIGH)

HIGH Z
(DISCONNECTED)
(CONTROL LINE
LOW)

(B) The three possible states for a tri-state buffer.

Fig. 12-4. Digital sections are switched into or out of a data bus by
means of tri-state buffers.

ment and data communication equipment. In turn, a line receiver
is used to convert the EIA signal levels to TTL logic levels.

Line Receiver Function

An EIA RS-232-C line receiver is basically an inverting level
shifter that inputs signals from an RS-232-C driver and outputs TTL
compatible logic levels. Three input channels are provided: conventional, hysteresis (slicing), and response (control) terminals.

Application of Logic Comparator

A logic comparator is not applicable for testing devices that
employ TTL input and MOS output, or MOS input and TTL output.
However, most logic probes and pulsers provide a choice of

Table 12-1. Bus Troubleshooting Tips

Type of Bus Driver	Troubleshooting Tips
Open Collector (Wired-AND/OR)	1. Open collectors can sink current, but not source it so a pull-up resistor to V_{cc} is connected to the output. 2. Disable driver input(s) 3. Pulse output(s) 4. Faulty driver will draw the most current
Single Driver	1. Driver can both source and sink current 2. Pulse input(s) 3. Probe output(s) for logic state changes, **or** 4. Current Trace output(s) for amplitude and the direction of the current path 5. Determine if driver is dead or bus is stuck 6. Replace dead driver, **or** 7. Pulse and Current Trace at output to pinpoint bus fault
Three State Buffer (With source and sink capability)	1. Disable driver input(s) 2. Pulse bus output lines 3. If one output draws current, verify if it is faulty **or** 1. Enable drivers 2. Pulse driver inputs individually 3. If one output fails to indicate current flow, verify if it is open

Courtesy Hewlett-Packard Co.

Note: A "stuck-at" bus trouble symptom can occur due to an open circuit at any one of three points: the bus driver output, the bus line itself, or a listener (data destination) input. Usually, the open circuit can be pinpointed by pulsing and probing, or by using a current tracer to check for presence or absence of current at inputs or outputs.

either TTL or MOS modes of operation. A current tracer is equally useful for tracking down faults in TTL or in MOS circuitry.

True/Complement Zero/One Element Function

A true/complement zero/one element is a type of controlled inverter that functions either to pass a binary number unchanged, or to complement the number in passage. For example, 1011 may be inputted and in turn outputted as 1011 or as 0100. The 1011 output is an example of the *true* mode of operation, whereas the 0110 output exemplifies the *complementary* mode of operation. A true/complement function is basic in arithmetic-logic units (alu's), inasmuch as addition occurs when the data is true, and subtraction

146

occurs when the data is complementary. Note that all arithmetical operations are performed by adders in combination with true/complement elements and various types of registers, such as shift registers.

Section 13
MEMORIES

A memory is defined as the equipment and media used to store machine-language information. Generally, the term *memory* denotes storage within a control system, whereas *storage* is used to refer to magnetic drums, disks, MOS devices, tapes, punched cards, and so on, which are external to the control system of a digital computer. Either term means to collect and to hold pertinent digital information until it is needed by the computer. There is no sharp dividing line between a shift register and a memory. When a shift register is operated as a memory, data bits may be clocked into and out of the device sequentially, or data words may be inputted or outputted in parallel. The various stages within a shift register are termed memory cells; the output may be connected back to the input, and the arrangement is then called an accumulator.

Shift-register memories may be very elaborate and may accommodate a kilobit of data, or more (a kilobit is equal to 1024 bits). There are two basic types of shift registers, termed dynamic and static designs. A dynamic memory employs memory cells that store data bits in capacitive elements. In turn, a dynamic memory must be clocked continuously at some minimum specified clock rate in order to refresh the charges in the capacitors. On the other hand, a static shift-register memory utilizes flip-flop type cells, and continuous clocking is not required. Simple registers are commonly included in the input lines and the output lines of large memories.

A widely used form of memory accepts 1 and 0 bits or digital words; these bits or words can be accessed in turn and outputted as required. A random-access memory, also termed a read-and-write memory, or RAM, may have data written into it from a keyboard, for example. Then, the data stored in the memory may be arbitrarily accessed and read out for performance of a chosen operation, such as addition. Another widely used type of memory contains 1 and 0 bits in permanent storage; it is called a read-only memory, or ROM, and is typically designed as a diode-logic array. Although the contents of a RAM can be erased, and new data can then be written into its cells, this is not possible with a conventional ROM. A ROM is programmed, which means that data can be written into the ROM only once; programming is commonly accomplished at the time of manufacture. As an illustration, an algorithm may be programmed into a ROM; then, whenever this algorithm is required in an arithmetic operation, it is read out of the ROM. Note also that the contents of an ordinary RAM are erased when its power supply is turned off; this is called a *volatile* form of memory. On the other hand, all ROMs are nonvolatile.

Not all ROMs are programmed at the time of manufacture. Thus a programmable read-only memory, or PROM, contains cells designed as transistors with fusible emitter resistors. At the outset, each cell represents a 0. To program a 1 into a given cell, an excessive amount of voltage is applied to the transistor, and the

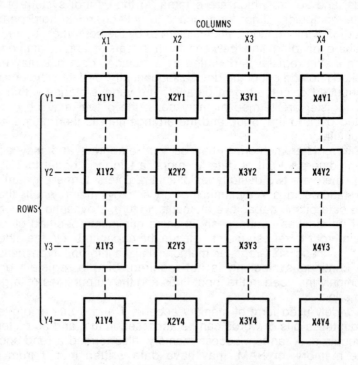

Fig. 13-1. Rows and columns of memory cells.

resulting abnormal current demand blows the fusible emitter resistor, so that a 1 is then stored in the particular cell. A more elaborate type of PROM is called an EPROM, or erasable programmable read-only memory. This specialized design of ROM can be erased and reprogrammed many times. A stored program is erased by exposure of the silicon chip to ultraviolet light. It is reprogrammed in the same manner as a PROM.

A widely used basic type of memory consists of rows and columns of flip-flops, as depicted in Fig. 13-1. Individual rows are usually designated as Y1, Y2, Y3, Y4, etc.; whereas, individual columns are designated as X1, X2, X3, X4, and so on. An *address* is a digital word that identifies a specific location in the memory. As an illustration, the arrangement in Fig. 13-1 has a capacity of 16 bits. If we stipulate that the unshaded blocks (FFs) represent 0 bits, then the shaded block represents a 1 bit with the address X2Y2. In the case that this diagram is for a RAM, it is indicated that a 1 bit has been written into the memory at X2Y2. Conversely, this 1 bit can be retrieved at X2Y2. Inasmuch as this is a RAM, the 1 at X2Y2 can be erased and replaced with a 0. Note that a RAM with a capacity of one kilobit is contained in an IC package with 16 pins.

Section 14
MISCELLANEOUS DIGITAL TESTS

Replacement pc boards may be available for modular digital systems. In such a case, the troubleshooter should start by substituting a known good board for a suspected defective board. It is sometimes helpful to "wiggle" a suspected board while monitoring the output with a logic probe in order to identify possible poor contacts.

Documentation Service Aids

Instruction manuals, maintenance manuals, and service indexes are often available in addition to circuit schematics and logic documentation. Manuals and service indexes will often provide facts concerning normal functions versus fault conditions. An instruction manual is primarily concerned with normal operation of the digital equipment. A maintenance manual provides timing diagrams, often with photographs of the pertinent oscilloscope patterns. A service index includes charts for troubleshooting procedures. Unless the technician is familiar with the equipment under test, it is essential that he or she consult all available documentation.

Localizing an Open Circuit

An open signal path (break in a pc conductor) is not indicated as a fault by a logic comparator; the ICs will all pass the comparator test. When an open signal path is suspected, the logic pulser and probe should be applied to check for continuity. Thereby, the technician can determine the precise point of discontinuity.

Use of "Smart Scopes"

Various professional oscilloscopes designed for digital analysis are in the "intelligent oscilloscope" or "smart scope" category. These oscilloscopes include microprocessor circuitry to provide LED or on-screen readout of parameters such as peak-to-peak voltage, frequency, elapsed time between operator-selected points on waveforms, sweep speed, and so on. The illustrated "smart scope" in Fig. 14-1 features delta time measurement. It incorporates a system of two intensified markers with LED readout—automatic calculation of rise time, propagation delay, and clock phasing are provided, with very high accuracy. In addition, this type of oscilloscope includes dmm facilities, whereby readout is provided of ac voltage, dc voltage, dc current, or resistance that may be present between the vertical-input test prods.

Fig. 14-1. A 275 MHz professional "intelligent oscilloscope."
(Courtesy Hewlett-Packard Co.)

Part 2

DIGITAL

TROUBLESHOOTING

TESTS

Section 15

GATE TESTS

15-1 To Check a Multiple-Input AND Gate

Equipment: Logic probe and logic pulser.

Connections Required: With reference to Figs. 15-1 and 2, connect the probe power leads and the pulser power leads to the power supply in the circuit under test, or to an external power supply or battery with suitable voltage (for example, TTL gates operate from power supplies in the range from 4.5 to 15 volts). In the second part of the test, connect the inputs of the AND gate together (a multipin stimulus cable assembly provides convenience in connecting IC pins together).

Procedure: Apply the logic probe at the output of the AND gate. Apply the logic pulser in turn to each input of the AND gate, and

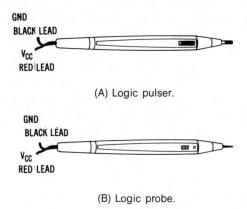

(A) Logic pulser.

(B) Logic probe.

Fig. 15-1. The two most basic digital logic test instruments.

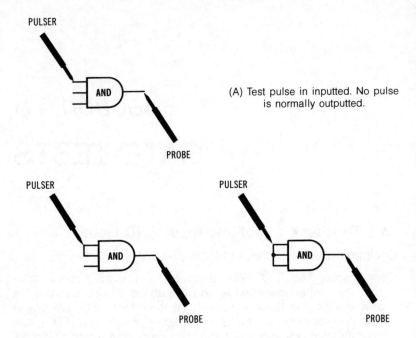

(A) Test pulse in inputted. No pulse is normally outputted.

(B) Test pulse is inputted. No pulse is normally outputted.

(C) Test pulse is inputted. Pulse is normally outputted.

Fig. 15-2. Pulse injection tests are made.

drive the input with a test pulse, observing the probe indication. Repeat this test with two input pins connected together; then repeat the test with three input pins connected together. Continue testing until the AND gate is tested with all of its input pins connected together (see Fig. 15-3).

Evaluation of Results: An AND gate normally produces no output unless the test pulse is applied with all of its input pins connected together (see Fig. 15-4). An output indication will normally be obtained when all of the inputs are pulsed simultaneously. Otherwise, the AND gate is defective and should be replaced. *Tests are ordinarily made in-circuit. Out of circuit tests require a bench power supply or battery.*

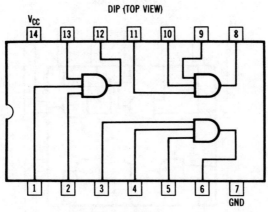

DIP (TOP VIEW)

Positive logic: **Y = ABC**

(A) Triple 3-input AND gate.

DIP (TOP VIEW)

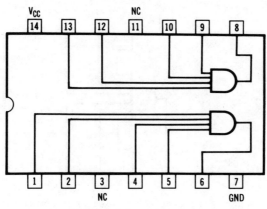

Positive logic: **Y = ABCD**

(B) Dual 4-input AND gate. *(Courtesy Fairchild Camera and Instrument Corp.)*

Fig. 15-3. Typical packages and pinouts.

159

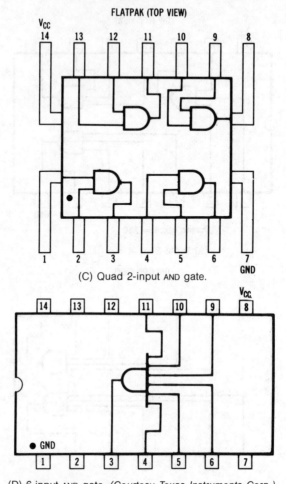

FLATPAK (TOP VIEW)

(C) Quad 2-input AND gate.

(D) 6-input AND gate. *(Courtesy Texas Instruments Corp.)*

Fig. 15-3. Typical packages and pinouts. (Continued)

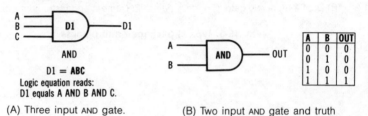

(A) Three input AND gate.

D1 = **ABC**
Logic equation reads:
D1 equals A AND B AND C.

A	B	OUT
0	0	0
0	1	0
1	0	0
1	1	1

(B) Two input AND gate and truth table.

Fig. 15-4. AND gates, logic equation, and truth table.

15-2 To Check a Multiple-Input OR Gate

Equipment: Same as in 15-1.

Connections Required: Same as in 15-1.

Procedure: Same as in 15-1.

Evaluation of Results: An OR gate normally produces a high output when the test pulse is applied to any one or to all of its inputs, when its inputs are initially low. In case that its inputs are initially high, it normally produces a low output only when all of its inputs are driven low. Otherwise, the OR gate is defective and should be replaced. Tests are ordinarily made in-circuit. Out of circuit tests require a bench power supply or battery (see Figs. 15-5 and 6).

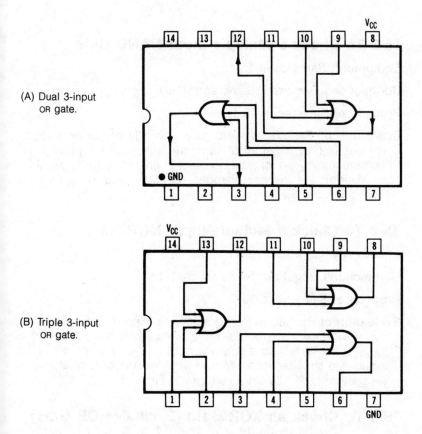

(A) Dual 3-input OR gate.

(B) Triple 3-input OR gate.

Fig. 15-5. Small scale integration OR gate packages.

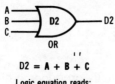

$$D2 = A + B + C$$

Logic equation reads:
D2 equals A OR B OR C.

Fig. 15-6. Symbol, logic equation, and truth table for OR gate.

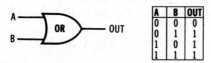

A	B	OUT
0	0	0
0	1	1
1	0	1
1	1	1

15-3 To Check a Multiple-Input NAND Gate

Equipment: Same as in 15-1.

Connections Required: Same as in 15-1.

Procedure: Same as in 15-1.

Evaluation of Results: A NAND gate normally produces a logic-low output only when all of its inputs are pulsed logic high simultaneously. A NAND gate consists of an AND gate followed by an inverter; the inverter changes 0s to 1s, and 1s to 0s (see Figs. 15-7 and 8).

15-4 To Check a Multiple-Input NOR Gate

Equipment: Same as in 15-1.

Connections Required: Same as in 15-1.

Procedure: Same as in 15-1.

Evaluation of Results: A NOR gate normally produces a logic-low output whenever any one or more of its inputs are pulsed logic high; it normally has a logic-high output only when all of its inputs are simultaneously logic low. A NOR gate consists of an OR gate followed by an inverter (see Figs. 15-9 and 10).

15-5 To Check an XOR Gate (Exclusive OR Gate)

Equipment: Logic pulser and logic probe.

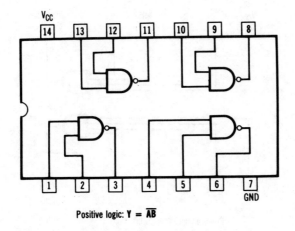

Positive logic: Y = $\overline{AB}$

(A) Quad 2-input gate.

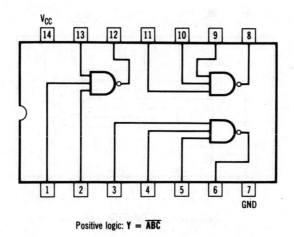

Positive logic: Y = $\overline{ABC}$

(B) Triple 3-input gate.

Fig. 15-7. Typical NAND gate packages and pinouts.

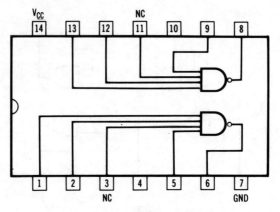

Positive logic: $Y = \overline{ABCD}$

(C) Dual 4-input gate.

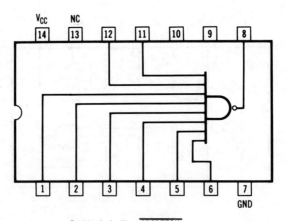

Positive logic: $Y = \overline{ABCDEFGH}$

(D) 8-input gate.

Fig. 15-7. Typical NAND gate packages and pinouts. (Continued)

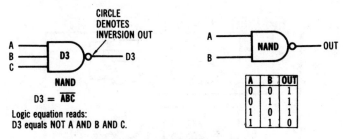

CIRCLE DENOTES INVERSION

NAND

$D3 = \overline{ABC}$

Logic equation reads:
D3 equals NOT A AND B AND C.

Fig. 15-8. Symbol, logic equation, and truth table for NAND gate.

Connections Required: Connect the V_{cc} and gnd leads of the pulser and probe to the power supply in the equipment under test. Apply the probe at the output of the XOR gate. Apply the pulser to one of the gate inputs; then apply the pulser to both of the gate inputs.

Procedure: Inject a test pulse first into one of the input leads, and then into both of the input leads. Observe the gate output response with the logic probe.

Evaluation of Results: An output pulse should be produced by the XOR gate whenever a test pulse is injected into one of the gate inputs. On the other hand, no output pulse should be produced when the test pulse is applied simultaneously to both of the XOR gate inputs. Failure to respond normally to these tests indicates that the XOR gate is defective (see Fig. 15-11).

15-6 To Check an AND-OR-INVERT Gate

Equipment: Logic pulser and logic probe.

Connections Required: Connect the V_{cc} and gnd leads of the pulser and probe to the power supply in the equipment under test. Apply the probe at the output of the device. Apply the pulser to the AND gate inputs, as explained next.

Procedure: Inject a test pulse in turn at each individual AND-gate input, and observe probe indication. Then, inject a test pulse in turn at each set of AND-gate inputs, and observe probe indication.

Evaluation of Results: In normal operation, no output is obtained from the device unless the test signal is applied simultaneously

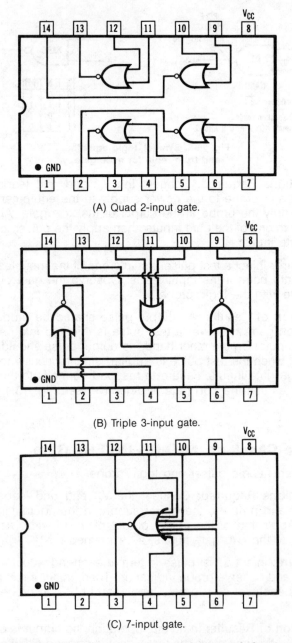

(A) Quad 2-input gate.

(B) Triple 3-input gate.

(C) 7-input gate.

Fig. 15-9. Examples of NOR gate packages and pinouts.

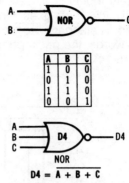

Logic equation reads:
D4 equals NOT A OR B OR C.

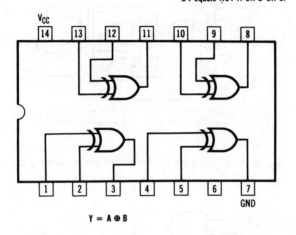

$$Y = A \oplus B$$

Fig. 15-11. Typical XOR gate package and pinout, with logic equation.

to a set of AND-gate inputs. Otherwise, the AND-OR-INVERT gate is defective and should be replaced (see Figs. 15-12 and 15-13).

15-7 To Check an XNOR (Exclusive NOR) Gate

Equipment: Same as in 15-5.

Connections Required: Same as in 15-5.

Procedure: Same as in 15-5.

Evaluation of Results: Same as in 15-5; the only distinction is that the $\overline{\text{XOR}}$ output is initially high, and goes low when either one of the gate inputs is pulsed. On the other hand, the XOR output is initially low, and goes high when either of the gate inputs is pulsed. A standard $\overline{\text{XOR}}$ package also provides XOR outputs. In other words, the package provides complementary outputs (see Fig. 15-14).

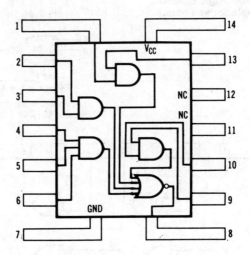

Fig. 15-12. A four-wide 2-2-2-3-input AND-OR-INVERT gate.

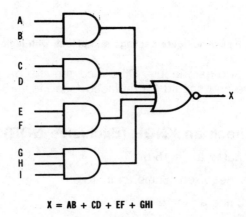

X = AB + CD + EF + GHI

Fig. 15-13. AND-OR-INVERT gate symbol and logic equation.

168

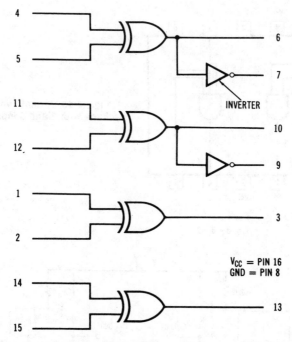

Fig. 15-14. Logic diagram with pinouts for xor gate ($\overline{XOR}$ outputs are provided at pins 7 and 9).

15-8 To Check a Dual AND Gate with V_{CC} and V_{EE} Power Supplies

Equipment: Logic probe and logic pulser.

Connections Required: Connect the V_{CC} and gnd leads of the pulser and probe to an *external* battery (5 V battery in this example). Connect the negative terminal of the battery to the gnd bus in the equipment under test (see Figs. 15-15 and 16).

Procedure: Same as in 15-1.

Evaluation of Results: Same as in 15-1.

15-9 To Trace a V_{CC}-Gnd Short External to the IC Packages

Equipment: Logic pulser and logic current tracer.

Connections Required: Connect V_{CC} and gnd leads of pulser and tracer to an external 5-volt supply.

169

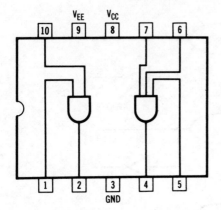

Fig. 15-15. A dual AND gate
package with 2 and 3 input gates.

Positive logic: 2 = 1 · 10
4 = 5 · 6 · 7

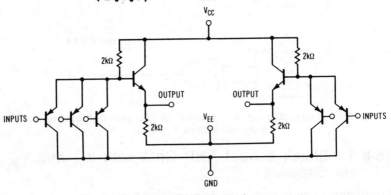

Fig. 15-16. Schematic diagram of an AND gate.

Procedure: Remove power from the circuit under test. Lift one side of the electrolytics on the supply bus (this speeds up troubleshooting by a factor of 10 because electrolytics "eat" pulses and produce confusing current paths). Inject test pulses across the power supply pins, or across components in the corners. Moving the pulsing point around from corner to corner and tracing current flow from the pulsing point with the current probe helps to speed up fault location. Because the test pulses flow into a short, ample current is available; set tracer sensitivity to 1 amp (see Fig. 15-17).

Evaluation of Results: The indicator lamp in the current tracer glows in proportion to the intensity of the current flowing at or near the tracer tip. As you pass the short, the indicator lamp will

170

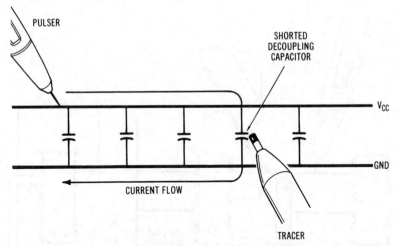

Fig. 15-17. A V_{cc} to gnd short may be traced to a shorted decoupling capacitor. *(Courtesy Hewlett-Packard Co.)*

go dark. When it appears that you have located the short, verify it by moving the pulse-injection point *to* the short. In turn, no current paths will be detected elsewhere on the pc board as you are actually injecting the pulse directly across the short.

15-10 To Trace Solder/Gold/Copper Bridge Faults

Equipment: Logic pulser and current tracer.

Connections Required: Connect V_{cc} and gnd leads of pulser and tracer to power supply in the equipment under test. (However, if electrolytics may be encountered, follow the 15-9 test procedure.)

Procedure: Use the pulser to pulse the driver output on the faulty node at a desired pulse rate. Adjust the sensitivity of the current tracer at the node driver output; then use the current tracer to track down the current pulses to the short (see Fig. 15-18).

Evaluation of Results: The indicator light on the current tracer will go out when the tracer tip passes the solder bridge.

15-11 To Troubleshoot a "Stuck-At" Node

Equipment: Same as in 15-10.

Connections Required: Same as in 15-10.

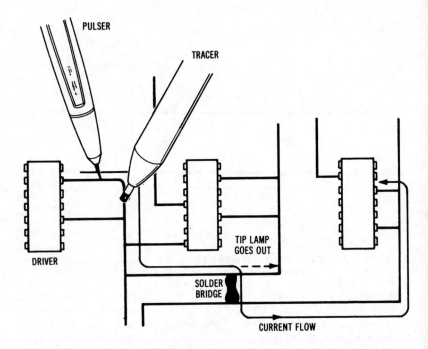

Fig. 15-18. Tracing pc conductors to a solder/gold/copper bridge fault.
(Courtesy Hewlett-Packard Co.)

Procedure: The technician needs to determine whether the driver is dead, or whether a shorted input is clamping the node to a fixed value. The logic probe and pulser are applied to test the logic state of the node and to observe whether this state can be changed. For example, shorts to V_{cc} or to gnd cannot be overridden by pulsing. Injection of pulses at the node enables the technician to follow the current flow directly to the faulty input. Tests are facilitated by adjusting the tracer sensitivity to a point that the indicator light is just visible with the pulser set to its 100-Hz mode.

Evaluation of Results: When the indicator light suddenly becomes dark, the tracer tip has passed the fault point (see Fig. 15-19).

15-12 To Check an IC That Has an Apparent Internal Short

Equipment: Same as in 15-10.

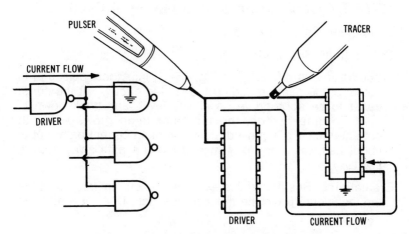

Fig. 15-19. Troubleshooting a "stuck-at" node.
(Courtesy Hewlett-Packard Co.)

Connections Required: Same as in 15-10.

Procedure: The NOR gate depicted in Fig. 15-20 tests bad when checked with a logic probe and logic pulser. Hence, the technician's problem is to determine the nature of the fault before removing the apparently internally shorted IC. First, a test pulse should be injected at pin 12; it will be observed with the probe that pin 13 changes state, but changes in the wrong direction (pins 12 and 13 are in the *same* state during test). Therefore, the technician should pulse pin 12 and check current flow at pin 13 with a current tracer; then, the pulser and current tracer should be reversed. In this type of fault, it will be found that the current flow is identical in both tests.

Evaluation of Results: The foregoing test results indicate that pins 12 and 13 are shorted together. In a case history, the short was determined to be caused by a solder bridge on the back of the circuit board.

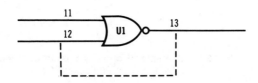

Fig. 15-20. Checking a digital IC that has an apparent internal short.
(Courtesy Hewlett-Packard Co.)

173

15-13 To Track Down a Multiple-Input Fault

Equipment: Same as in 15-10.

Connections Required: Same as in 15-10.

Procedure: With reference to Fig. 15-21, inject test pulses at the output of the gate preceding the faulted circuit. Set the current tracer to an appropriate sensitivity level, and track down the fault. In other words, the indicator lamp glows while the current-tracer tip is moved along the main current path, but extinguishes when the tip is moved along a path away from the short.

Evaluation of Results: In this example, the current tracer will lead the technician up to the input of NOR gate U5A, beyond which

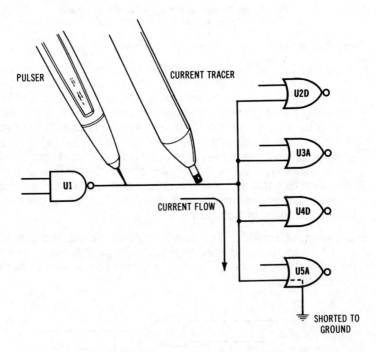

Fig. 15-21. Example of a multiple input fault (short).
(Courtesy Hewlett-Packard Co.)

the circuit enters the gate. Therefore, it is concluded that U5A has an internal short to ground.

174

15-14 To Troubleshoot Wired-AND/Wired-OR Circuits

Equipment: Same as in 15-10.

Connections Required: Same as in 15-10.

Procedure: With reference to Figs. 15-22 through 25, the open collectors of several NAND gates are connected together and to a common external load resistor. If any one of the gates goes low, it controls the bus, because an open-collector gate cannot supply any current in a high logic state. In turn, if any NAND gate becomes "stuck low," all of the other gates also appear to be stuck low. If all of the gates are driven high, only the stuck-low gate will draw current from R_L. Hence, a current-tracing probe is utilized.

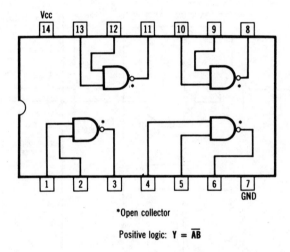

*Open collector

Positive logic: $Y = \overline{AB}$

Fig. 15-22. Typical NAND gate package with open collector outputs.

Evaluation of Results: When the branches of the common interconnection node are current-traced, the faulty gate is identified as the one that draws current from R_L while all of the gates are being driven high.

15-15 To Troubleshoot a Wired-AND/Wired-OR Configuration With 4-2-3-2-Input AND-OR-INVERT Gates

Equipment: Same as in 15-10.

175

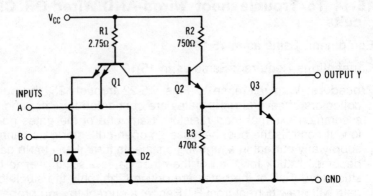

Fig. 15-23. Schematic for NAND gate with open collector output.

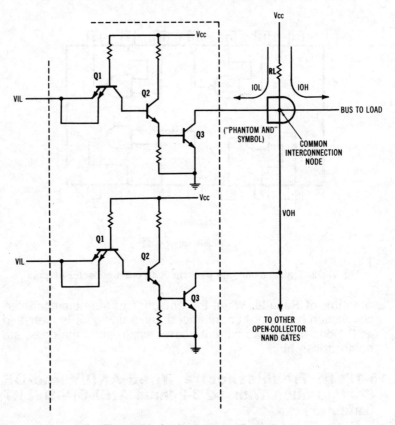

Fig. 15-24. A wired-AND configuration.

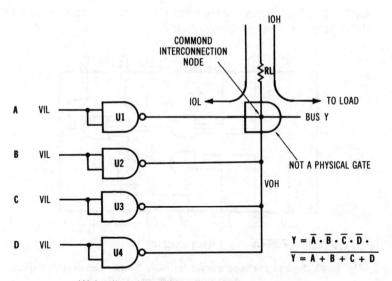

(A) Implied dot wired-AND or wired-OR connection.

$$Y = \overline{A} \cdot \overline{B} \cdot \overline{C} \cdot \overline{D} \cdot$$
$$\overline{Y = A + B + C + D}$$

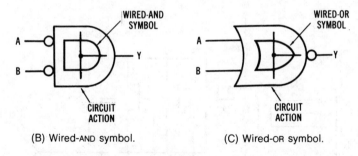

(B) Wired-AND symbol.　　　　　(C) Wired-OR symbol.

Fig. 15-25. Wired-AND and wired-OR symbol and connection.

Connections Required: Same as in 15-14.

Procedure: With reference to Figs. 15-26 and 27, the AND-OR-INVERT gate operates as a wired-OR configuration, with the open collector NOR gate serving as the common interconnection node. Two or more gates may have their open collectors connected together, if desired, with a common external load resistor. If all of the outputs are driven high, a stuck-low gate will draw current from the load resistor. Hence, a current-tracing probe is utilized.

Evaluation of Results: When the branches of the common interconnection node are current-traced, the faulty gate is identified as the one that draws current from the load resistor while all of the outputs are being driven high.

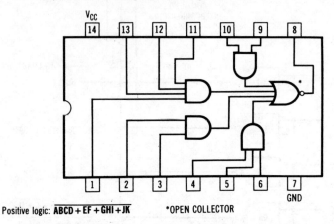

Positive logic: $\overline{ABCD + EF + GHI + JK}$ *OPEN COLLECTOR

Fig. 15-26. Typical package pinout for 4-2-3-2-input AND-OR-INVERT gate with open collector.

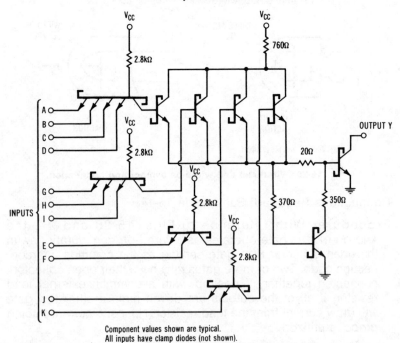

Component values shown are typical.
All inputs have clamp diodes (not shown).

Fig. 15-27. Schematic diagram of the 4-2-3-2-input AND-OR-INVERT gate with open collector.

15-16 To Estimate HIGH and LOW State Current Values

Equipment: Logic pulser and current tracer.

Connections Required: Connect V_{cc} and gnd leads of pulser and tracer to the power supply in the equipment under test.

Procedure: With reference to Fig. 15-28, pulse the circuit high and low. Estimate current levels from the response of the current tracer.

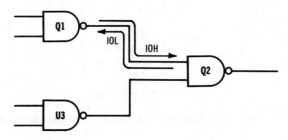

Fig. 15-28. TTL circuitry. *(Courtesy Hewlett-Packard Co.)*

Evaluation of Results: Check current-tracer adjustment and estimate the current levels accordingly. In the following diagram, I_{OH} denotes high-level output current (logic 1); in TTL, this is approximately 40 μA. I_{OL} denotes low-level output current (logic 0); in TTL, this is normally about 1.6 mA for a fan-out of one (a single load). However, if the load is shorted, TTL low-level output current may be as high as 55 mA. Partial shorts result in abnormal pulse currents less than 55 mA.

15-17 Oscilloscope Check of AND Gate Operation

Equipment: Pulse generator, oscilloscope, and suitable battery to provide V_{cc} voltage for the AND gate.

Connections Required: As shown in Fig. 15-29, the battery is connected in suitable polarity to the V_{cc} and gnd terminals of the gate; the output from the pulse generator is connected to the various inputs of the gate; the scope is connected to the output of the gate.

Procedure: Use correct polarity of pulse, and suitable amplitude; pulse width and repetition rate must be within the device ratings. Apply pulse voltage in turn to each of the gate inputs, to pairs of the inputs, and finally to all of the inputs simultaneously.

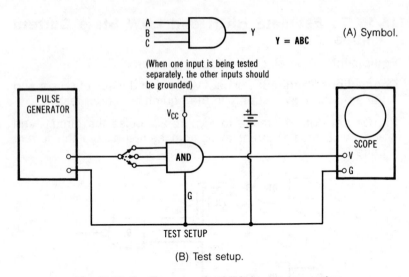

(A) Symbol.

$$Y = ABC$$

(When one input is being tested separately, the other inputs should be grounded)

(B) Test setup.

Fig. 15-29. Oscilloscope check of AND gate operation.

Evaluation of Results: In normal operation, no output will be obtained from the AND gate unless all of its inputs are simultaneously driven logic-high. *This is an out-of-circuit test; for an in-circuit test, see 15-18.*

15-18 In-Circuit Check of AND Gate Operation With Oscilloscope

Equipment: Oscilloscope, and any suitable arrangement for application of different digital words to the AND-gate inputs. The digital words should be repeated for the duration of the test, as by appropriate programming of a digital computer.

Connections Required: If a single-trace scope is used, the output from the gate is connected to the vertical-input channel. If a dual-trace scope is used, the vertical-input channels are connected to two inputs, to one input and to the output, or to the other input and the output of the gate (see Figs. 15-30 and 31).

Procedure: With the gate being driven as in normal operation, observe the output waveform; one of the input waveforms may be compared with the output waveform if a dual-trace scope is used. Pairs of input waveforms may also be compared with a dual-trace scope.

Evaluation of Results: In normal operation, an output pulse is ob-

180

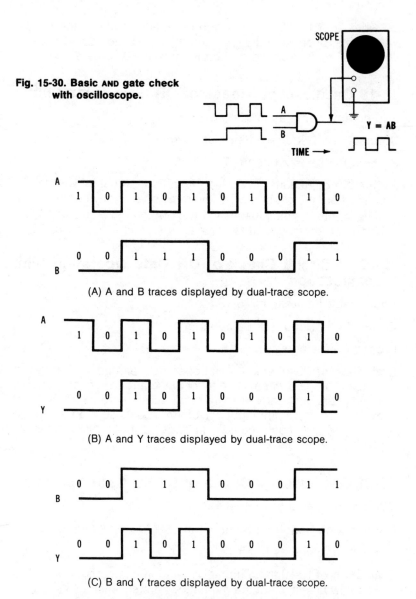

Fig. 15-30. Basic AND gate check with oscilloscope.

SCOPE

A

Y = AB

B

TIME →

A

1 0 1 0 1 0 1 0 1 0

B

0 0 1 1 1 0 0 0 1 1

(A) A and B traces displayed by dual-trace scope.

A

1 0 1 0 1 0 1 0 1 0

Y

0 0 1 0 1 0 0 0 1 0

(B) A and Y traces displayed by dual-trace scope.

B

0 0 1 1 1 0 0 0 1 1

Y

0 0 1 0 1 0 0 0 1 0

(C) B and Y traces displayed by dual-trace scope.

Fig. 15-31. Dual-trace displays of AND gate operation.

tained only when all of the inputs are simultaneously driven logic-high. In the accompanying example, the input digital words are 1010101010 and 0011100011; the normal output digital word is 0010100010. It is good practice to check the input waveforms, because a defect in the driving section could cause "garbage" to be applied to the inputs of the AND gate.

15-19 Oscilloscope Check of OR Gate Operation

Equipment: Same as in 15-17.

Connections Required: Same as in 15-17.

Procedure: Same as in 15-17.

Evaluation of Results: In normal operation, an output will be obtained from the OR gate whenever any one, or any combination, of its inputs is simultaneously driven logic-high (see Fig. 15-32). *This is an out-of-circuit test; for an in-circuit test, see 15-20.*

15-20 In-Circuit Check of OR Gate Operation With Oscilloscope

Equipment: Same as in 15-18.

Connections Required: Same as in 15-18.

Procedure: Same as in 15-18.

Evaluation of Results: In normal operation, an output is obtained from the OR gate when any one, or any combination, of its inputs is simultaneously driven logic-high (see Fig. 15-33). It is good practice to check the input waveforms, because a defect in the driving section could cause "garbage" to be applied to the inputs of the OR gate.

15-21 Oscilloscope Check of NAND Gate Operation

Equipment: Same as in 15-17.

Connections Required: Same as in 15-17.

Procedure: Same as in 15-17.

Evaluation of Results: In normal operation, the output will not go logic-low unless all of the gate inputs are simultaneously driven logic-high. *This is an out-of-circuit test; for an in-circuit test, see 15-22.*

15-22 In-Circuit Check of NAND Gate Operation With Oscilloscope

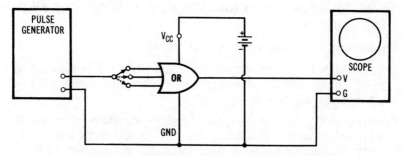

Fig. 15-32. Oscilloscope check of OR gate operation.

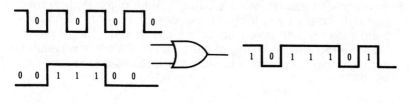

TIME ——▶

Fig. 15-33. Typical input and output waveforms displayed during an
in-circuit check of an OR gate.

Equipment: Same as in 15-18.

Connections Required: Same as in 15-18.

Procedure: Same as in 15-18.

Evaluation of Results: In normal operation, the output goes logic-low only when all of the NAND-gate inputs are simultaneously driven logic-high. For example, if the input digital words are 1010101010 and 0011100011, the normal output digital word is 1101011101 (see Fig. 15-34).

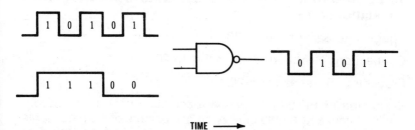

TIME ——▶

Fig. 15-34. Typical input and output waveforms observed during an
in-circuit check of a NAND gate.

15-23 To Track Down and Display a Glitch in a Digital System

Equipment: High-performance oscilloscope.

Connections Required: Apply low capacitance scope probe at points in the configuration where it is suspected that a glitch is causing a malfunction.

Procedure: Observe the digital waveform carefully, with ample trace brightness. A glitch may be very narrow compared to the width of a digital pulse. In turn, the glitch will be dim and can easily be overlooked.

Evaluation of Results: When a barely visible glitch is discerned in a pulse train, try to trigger the scope on the glitch. Then expand the glitch at increased brightness to verify its presence. Observe the relative timing of the glitch in the pulse waveform. In turn, it may be possible to reason back to its source (see Fig. 15-35).

15-24 Oscilloscope Check of NOR Gate Operation

Equipment: Same as in 15-19.

Connections Required: Same as in 15-19.

Procedure: Same as in 15-19.

Evaluation of Results: A logic-low output is normally obtained from a NOR gate whenever any one of its inputs is driven logic-high. A logic-low output should also be obtained whenever any combination of inputs is driven logic-high. *This is an out-of-circuit test; for an in-circuit test, see 15-25.*

15-25 In-Circuit Check of NOR Gate Operation With Oscilloscope

Equipment: Same as in 15-20.

Connections Required: Same as in 15-20.

Procedure: Same as in 15-20.

Evaluation of Results: In normal operation, a low output is obtained from a NOR gate only when one or more of the gate inputs are driven high. As exemplified in Fig. 15-36, if the NOR-gate inputs are 101010 and 111001, the normal output will be 000101.

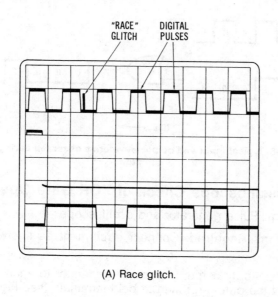

(A) Race glitch.

(B) Glitch is often invisible at low sweep speed.

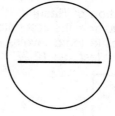

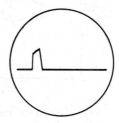

(C) Glitch becomes visible at higher sweep speed and increased brightness.

(D) Glitch is clearly visible and waveform is displayed at very high sweep speed and increased brightness.

Fig. 15-35. Typical race glitch displayed on screen of triple-trace lab-type scope.

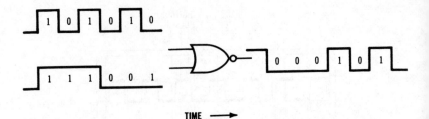

Fig. 15-36. Typical input and output waveforms observed during test of NOR gate.

15-26 Oscilloscope Check of XOR Gate Operation

Equipment: Pulse generator and oscilloscope.

Connections Required: Connect equipment as shown in Fig. 15-37.

Procedure: Apply output from pulse generator to each input terminal of the gate, and then to both terminals (see Fig. 15-37).

Evaluation of Results: In normal operation, a high output is obtained from the XOR gate only when its two input terminals are at opposite logic levels. *This is an out-of-circuit test; for an in-circuit test see 15-27.*

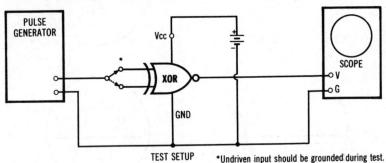

TEST SETUP *Undriven input should be grounded during test.

Fig. 15-37. Oscilloscope check of XOR gate operation.

15-27 In-Circuit Check of XOR Gate Operation With Oscilloscope

Equipment: Oscilloscope, and any suitable arrangement for application of different digital words to the XOR gate inputs. The digital words should be repeated for the duration of the test, as by appropriate programming of a digital computer.

Procedure: Check output waveform against the input waveforms.

Evaluation of Results: In normal operation, an XOR gate produces a high output only when both of its inputs are driven to opposite logic levels. Thus, if the input digital words are 0101010 and 0111000, the normal output word will be 001010 (see Fig. 15-38).

Connections Required: Apply low capacitance probe from scope to circuit point.

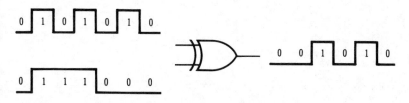

Fig. 15-38. Typical input and output waveforms observed during scope check of XOR gate operation.

15-28 Oscilloscope Check of XNOR Gate Operation

Equipment: Same as in 15-26.

Connections Required: Same as in 15-26.

Procedure: Same as in 15-26.

Evaluation of Results: In normal operation a low output is obtained from an XNOR gate only when its two input terminals are at opposite logic levels. *This is an out-of-circuit test; for an in-circuit test see 15-29.*

15-29 In-Circuit Check of XNOR Gate Operation

Equipment: Same as in 15-27.

Connections Required: Same as in 15-27.

Procedure: Same as in 15-27.

Evaluation of Results: In normal operation, a low output is obtained from an XNOR gate only when its two input terminals are driven to opposite logic levels. For example, if the input digital words are 0101010 and 0111000, the normal output word will be 110101.

Section 16
ADDERS

16-1 To Check a Dual Carry/Save Full Adder

Equipment: Logic pulser and logic probe.

Connections Required: Connect the V_{cc} and gnd leads of the pulser and probe to the power supply in the equipment under test. Connect a multistimulus test cable, or equivalent, to the various IC pins as required in the procedure.

Procedure: With reference to Fig. 16-1, pulse 1A, 1B, and $1C_n$ pins low; observe probe indication at output pin 1Σ and at $1C_{n+1}$. Each of the output pins normally has a low response. Repeat the test for input pin 1A pulsed high, with pin 1B and pin $1C_n$ pulsed low; output pin 1Σ will normally go high and output pin $1C_{n+1}$ will normally go low. Continue in similar manner until the truth table is verified for the first adder. Then repeat the series of tests to verify the truth table for the second adder.

Evaluation of Results: Unless each entry in the truth table is verified in the foregoing tests, it is indicated that the adder is defective and should be replaced.

16-2 To Check the Operation of a 2-Bit Full Adder

Equipment: Logic pulser and logic probe.

Connections Required: Connect the V_{cc} and gnd leads of the pulser and probe to the power supply in the equipment under test. Connect a multistimulus test cable, or equivalent, to the various pins as required in the procedure.

Procedure: With reference to Fig. 16-2, pulse A_1, B_1, C_{in}, A_2, and B_2 pins low; observe probe indication at output pins Σ_1, Σ_2, and

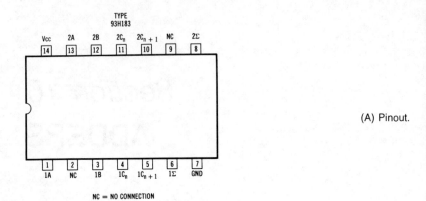

(A) Pinout.

NC = NO CONNECTION

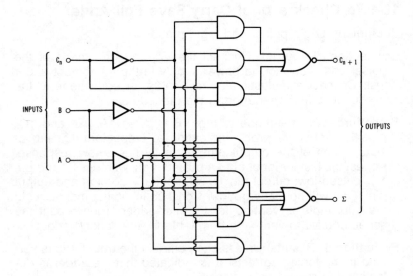

(B) Logic diagram. *(Courtesy Fairchild Camera and Instrument Corp.)*

INPUTS			OUTPUTS	
C_n	B	A	Σ	C_{n+1}
0	0	0	0	0
0	0	1	1	0
0	1	0	1	0
0	1	1	0	1
1	0	0	1	0
1	0	1	0	1
1	1	0	0	1
1	1	1	1	1

(C) Truth table.

Fig. 16-1. Typical dual carry/save full adder.

C_2. Repeat the test for A_1, B_1, A_2, and B_2 pins pulsed low, but with C_{in} pulsed high; observe the outputs produced at Σ_1, Σ_2, and C_2. Continue in the same manner until the truth table is completed.

Evaluation of Results: In normal operation, the adder will respond as specified in the truth table. For example, when inputs A_1, B_1, A_2, B_2, and C_{in} are low, Σ_1, Σ_2, and C_2 will normally be low. If the truth table is violated at any step in the test procedure, the adder is defective and should be replaced. *Note, however, that an externally shorted node could produce an incorrect response which would be falsely attributed to the adder.*

16-3 To Check the Operation of a 4-Bit Full Adder

Equipment: Same as in 16-2.

Connections Required: Same as in 16-2.

Procedure: Pulse the specified groups of inputs high or low (see Fig. 16-3). Observe the corresponding output indications by the logic probe at sum outputs Σ_1, Σ_2, Σ_3, Σ_4, and C_4.

Evaluation of Results: In normal operation, the adder will respond as specified in the truth table. *Note that Carry C_2 in the truth table is not pinned out in the IC package; instead, C_2 is an internal path.* If, for example, the test condition drives Σ_1 and Σ_2 to 11, C_2 will be low (0). On the other hand, one more input pulse will normally cause Σ_1 and Σ_2 to go to 00 with C_2 going high (1). In turn, Σ_3 will go high, and the readout will normally be 100.

16-4 To Check the Operation of a Dual Carry Dependent Full Adder

Equipment: Same as in 16-2.

Connections Required: Same as in 16-2.

Procedure: With reference to Fig. 16-4, check Adder 1 by first pulsing inputs A, B, and C low; check resulting outputs at $\overline{C}_o$, $\overline{S}$, and S with the logic probe. Next, tie inputs B and C to a fixed low level, and pulse input A high; observe the resulting outputs with the logic probe. Continue in a similar manner until the truth table for Adder 1 is completed. Then proceed to verify operation of Adder 2 with respect to its truth table.

Evaluation of Results: Unless each entry in the truth table for each adder is verified in the foregoing tests, it is indicated that the adder is defective and should be replaced. *Keep in mind, however, that sometimes an external short can make an IC "look bad."*

16-5 To Check the Operation of a Gated Full Adder

Equipment: Same as in 16-2.

Connections Required: Same as in 16-2.

Procedure: With reference to Fig. 16-5, first pulse inputs A, B, and C_n low; check resulting outputs at C_{n+1}, Σ, and $\overline{\Sigma}$ with the logic probe. Next, tie inputs B and C_n to a fixed low level, and pulse input A high; observe the resulting outputs with the logic probe. Continue in a similar manner until the truth table for the adder is completed.

Evaluation of Results: Unless each entry in the truth table is verified in the foregoing tests, it is indicated that the adder is defective and should be replaced.

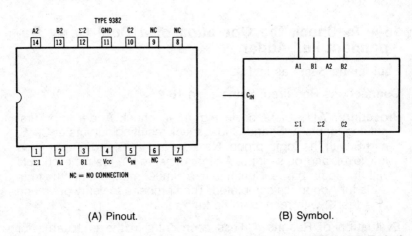

(A) Pinout.　　　　　(B) Symbol.

Fig. 16-2. Typical

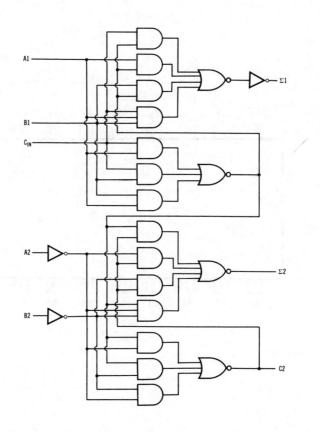

(C) Logic diagram. *(Courtesy Fairchild Camera and Instrument Corp.)*

(D) Truth table.

INPUT				OUTPUT					
				WHEN $C_{IN} = 0$			WHEN $C_{IN} = 1$		
A1	B1	A2	B2	Σ1	Σ2	C2	Σ1	Σ2	C2
0	0	0	0	0	0	0	1	0	0
1	0	0	0	1	0	0	0	1	0
0	1	0	0	1	0	0	0	1	0
1	1	0	0	0	1	0	1	1	0
0	0	1	0	0	1	0	1	1	0
1	0	1	0	1	1	0	0	0	1
0	1	1	0	1	1	0	0	0	1
1	1	1	0	0	0	1	1	0	1
0	0	0	1	0	1	0	1	1	0
1	0	0	1	1	1	0	0	0	1
0	1	0	1	1	1	0	0	0	1
1	1	0	1	0	0	1	1	0	1
0	0	1	1	0	0	1	1	0	1
1	0	1	1	1	0	1	0	1	1
0	1	1	1	1	0	1	0	1	1
1	1	1	1	0	1	1	1	1	1

2-bit binary adder.

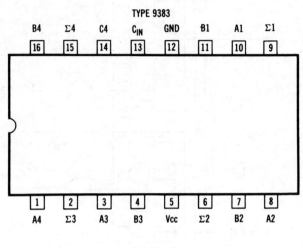

(A) Pinout.

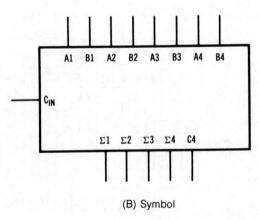

(B) Symbol

Fig. 16-3 Typical

194

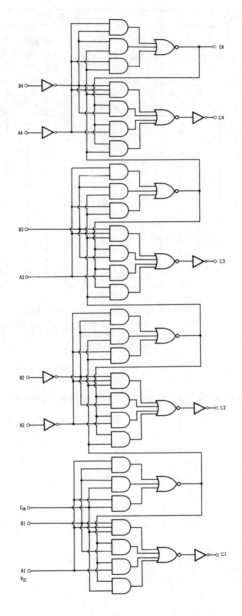

(C) Logic diagram. *(Courtesy Fairchild Camera and Instrument Corp.)*

4-bit binary adder.

INPUT				OUTPUT					
				WHEN $C_{IN}=0$ / WHEN $C2=0$			WHEN $C_{IN}=1$ / WHEN $C2=1$		
A1 / A3	B1 / B3	A2 / A4	B2 / B4	$\Sigma1$ / $\Sigma3$	$\Sigma2$ / $\Sigma4$	C2 / C4	$\Sigma1$ / $\Sigma3$	$\Sigma2$ / $\Sigma3$	C2 / C4
0	0	0	0	0	0	0	1	0	0
1	0	0	0	1	0	0	0	1	0
0	1	0	0	1	0	0	0	1	0
1	1	0	0	0	1	0	1	1	0
0	0	1	0	0	1	0	1	1	0
1	0	1	0	1	1	0	0	0	1
0	1	1	0	1	1	0	0	0	1
1	1	1	0	0	0	1	1	0	1
0	0	0	1	0	1	0	1	1	0
1	0	0	1	1	1	0	0	0	1
0	1	0	1	1	1	0	0	0	1
1	1	0	1	0	0	1	1	0	1
0	0	1	1	0	0	1	1	0	1
1	0	1	1	1	0	1	0	1	1
0	1	1	1	1	0	1	0	1	1
1	1	1	1	0	1	1	1	1	1

NOTE:
1. Input conditions at A1, A2, B1, B2 and C_{IN} are used to determine outputs $\Sigma1$ and $\Sigma2$, and the value of the internal carry C2. The values at C2, A3, B3 A4, and B4, are then used to determine outputs $\Sigma3$, $\Sigma4$ and C4.

(D) Truth table.

Fig. 16-3. Typical 4-bit binary adder. (Continued)

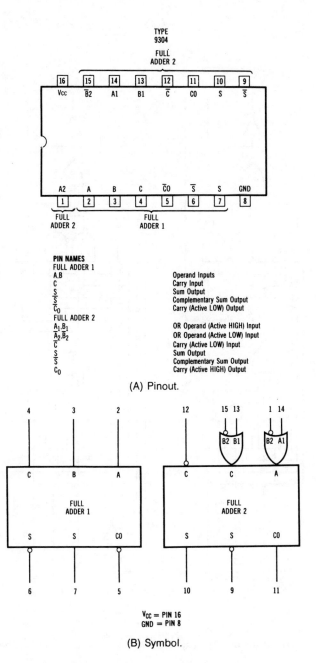

(A) Pinout.

(B) Symbol.

Fig. 16-4. Dual carry dependent full adder.

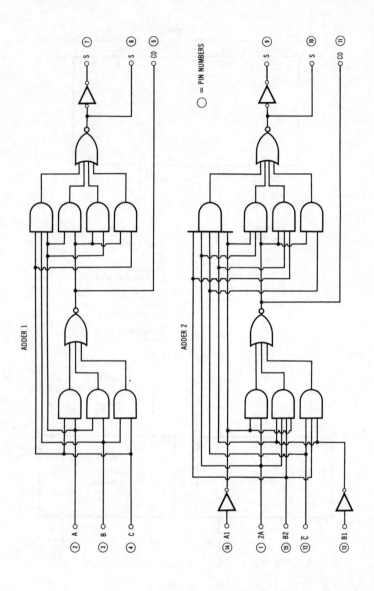

(C) Logic diagram. *(Courtesy Fairchild Camera and Instrument Corp.)*

Fig. 16-4. Dual carry dependent

INPUTS					OUTPUTS		
C	B1	A1	B̄2	Ā2	CO	S	S̄
0	0	0	0	0	1	1	0
0	0	0	0	1	1	0	1
0	0	0	1	0	1	0	1
0	0	0	1	1	0	1	0
0	0	1	0	0	1	1	0
0	0	1	0	1	1	1	0
0	0	1	1	0	1	0	1
0	0	1	1	1	1	0	1
0	1	0	0	0	1	1	0
0	1	0	0	1	1	0	1
0	1	0	1	0	1	1	0
0	1	0	1	1	1	0	1
0	1	1	0	0	1	1	0
0	1	1	0	1	1	1	0
0	1	1	1	0	1	1	0
0	1	1	1	1	1	1	0
1	0	0	0	0	1	0	1
1	0	0	0	1	0	1	0
1	0	0	1	0	0	1	0
1	0	0	1	1	0	0	1
1	0	1	0	0	1	0	1
1	0	1	0	1	1	0	1
1	0	1	1	0	0	1	0
1	0	1	1	1	0	1	0
1	1	0	0	0	1	0	1
1	1	0	0	1	0	1	0
1	1	0	1	0	1	0	1
1	1	0	1	1	0	1	0
1	1	1	0	0	1	0	1
1	1	1	0	1	1	0	1
1	1	1	1	0	1	0	1
1	1	1	1	1	1	0	1

(D) Adder 1 truth table.

INPUTS			OUTPUTS		
C	B	A	C̄O	S̄	S
0	0	0	1	1	0
0	0	1	1	0	1
0	1	0	1	0	1
0	1	1	0	1	0
1	0	0	1	0	1
1	0	1	0	1	0
1	1	0	0	1	0
1	1	1	0	0	1

(E) Adder 2 truth table.

full adder. (Continued)

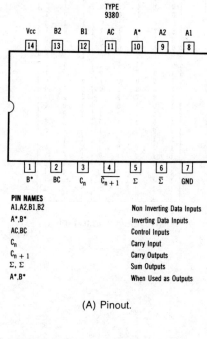

(A) Pinout.

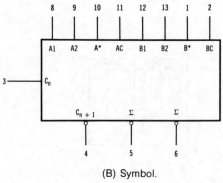

(B) Symbol.

Fig. 16-5. Typical

200

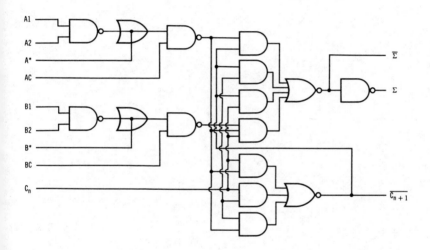

(C) Logic diagram. *(Courtesy Fairchild Camera and Instrument Corp.)*

C_n	B	A	$\overline{C_{n+1}}$	$\overline{\Sigma}$	Σ
0	0	0	1	1	0
0	0	1	1	0	1
0	1	0	1	0	1
0	1	1	0	1	0
1	0	0	1	0	1
1	0	1	0	1	0
1	1	0	0	1	0
1	1	1	0	0	1

NOTES:
1. $A = \overline{A^* \cdot AC}$, $B = \overline{B^* \cdot BC}$ where $A^* = \overline{A1 \cdot A2}$. $B^* = \overline{B1 \cdot B2}$
2. When A^* or B^* are used as inputs, A1 and A2 or B1 and B2 respectively must be connected to GND.
3. When A1 and A2 or B1 and B2 are used as inputs, A^* or B^* respectively must be open or used to perform Dot-OR logic.

(D) Truth table.

gated full adder.

LATCHES AND FLIP-FLOPS

17-1 To Check the Operation of a 4-Bit Latch

Equipment: Logic pulser and logic probe.

Connections Required: Connect V_{cc} and gnd leads of pulser and probe to power supply in the equipment under test.

Procedure: The clock input is driven from the equipment under test, as in usual operation. Test pulses are injected at the data input by the logic pulser. Resulting responses are observed at the Q output and at the $\overline{Q}$ output with the logic probe.

Evaluation of Results: The truth table stipulates that digital information present at a data (D) input will be transferred to the Q output when the clock goes high. The Q output level follows the data input level as long as the latter remains unchanged. Then, when the data input level changes, the information that was present at the data input when the clock went low will be retained briefly until the clock goes high. Thereupon, the changed data level will be transferred to the Q output (see Fig. 17-1).

17-2 To Check the Operation of a JK Master/Slave Flip-Flop

Equipment: Logic pulser and logic probe.

Connections Required: Tie the J and K terminals of one flip-flop together (see Fig. 17-2A). The R_D input should be tied to a fixed high logic level, or to an external 3-V battery. Connect V_{cc} and gnd leads of pulser and probe to power supply in the equipment under test.

Procedure: Inject pulses from the logic pulser into the paralleled JK terminals. Observe the output from Q and from $\overline{Q}$ with the logic probe. The clock terminal is activated from the equipment under test as in normal operation (see Fig. 17-2).

Evaluation of Results: Since the J and K inputs are tied together, the configuration operates as a *toggle* flip-flop; in other words, the Q and $\overline{Q}$ outputs will alternate between high and low levels on each clock pulse, as long as the JK tie is high; on the other hand, when the JK tie is low, the outputs will remain in the previous (clock pulse) state. Otherwise, the flip-flop is defective and should be replaced. Repeat the test for the second flip-flop.

17-3 To Check the Operation of a JK Master/Slave Flip-Flop With AND-OR Inputs

Equipment: Logic pulser and logic probe.

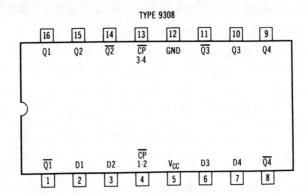

TYPE 9308

Positive logic: See truth table.
NC-No internal connection

(A) Pinout.

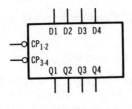

(B) Symbol.

Fig. 17-1. Typical

Connections Required: Tie J1A and J1B to each other and to K1A and K1B. The preset (S_D) input terminal should remain at a high logic level during the test; if necessary, S_D may be tied to a fixed high level or to an external 3-V battery. (A preset input serves to initialize the flip-flop by forcing Q to high level, thereby ensuring that all following logic sequences will proceed correctly.) Power the pulser and probe from the power supply in the equipment under test (see Fig. 17-3).

Procedure: Same as in 17-2.

Evaluation of Results: Same as in 17-2.

17-4 To Check the Operation of a JK Master/Slave Flip-Flop With AND Inputs

Equipment: Logic pulser and logic probe.

Connections Required: Tie J1, J2, and J3 together and to K1, K2, and K3. Preset (S_D) and clear (R_D) terminals should remain at a high logic level during the test; if necessary, S_D and R_D may be tied to a fixed high level, or to an external 3-V battery. Power the pulser and probe from the power supply in the equipment under test.

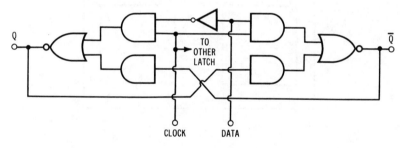

(C) Logic diagram. *(Courtesy Fairchild Camera and Instrument Corp.)*

(D) Truth table.

t_n	t_{n+1}
D	Q
1	1
0	0

NOTES:
t_n = bit time before clock negative-going transition.
t_{n+1} = bit time after clock negative-going transition.

4-bit latch.

Procedure: Same as in 17-1.

Evaluation of Results: Same as in 17-1 (see Fig. 17-4).

17-5 To Check a D-Type Edge-Triggered Flip-Flop

Equipment: Logic pulser and logic probe.

Connections Required: The preset and clear inputs should remain at a high level during the test; if necessary, they may be tied to a fixed high level, or to an external 3-V battery. Power the pulser and probe from the power supply in the equipment under test.

Procedure: Apply test pulses at the data (D) input, and check the Q and $\overline{Q}$ outputs with the logic probe. The clock terminal is activated from the equipment under test, as in normal operation.

Evaluation of Results: The input/output relations should occur in accordance with the truth table (see Fig. 17-5).

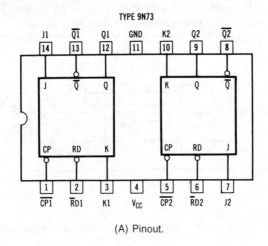

(A) Pinout.

(B) Symbol.

Fig. 17-2. Dual

17-6 To Check the Operation of a JK Master/Slave Flip-Flop With Separate Presets, Clears, and Clocks

Equipment: Logic pulser and logic probe.

Connections Required: Connect the V_{cc} and gnd leads of the pulser and probe to the power supply in the equipment under test. Tie the J and K inputs together. The preset and clear inputs should be at high level during the test; if necessary, they may be tied to a fixed high level, or to an external 3-V battery.

Procedure: Apply test pulses from the logic pulser to the paralleled J and K inputs. Check the Q and $\bar{Q}$ outputs with the logic probe. The clock terminal is activated from the equipment under test, as in normal operation.

Evaluation of Results: Same as in 17-2 (see Fig. 17-6).

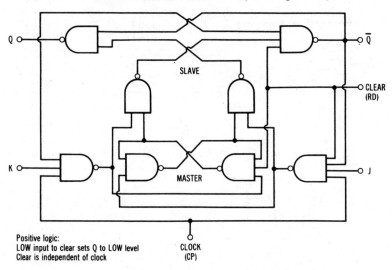

Positive logic:
LOW input to clear sets Q to LOW level
Clear is independent of clock

(C) Logic diagram. *(Courtesy Fairchild Camera and Instrument Corp.)*

(D) Truth table.

	t_n		t_{n+1}
J	K		Q
0	0		Q_n
0	1		0
1	0		1
1	1		Q_n

NOTES:
t_n = Bit time before clock pulse.
t_{n+1} = Bit time after clock pulse.

JK flip-flop.

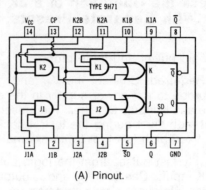

(A) Pinout.

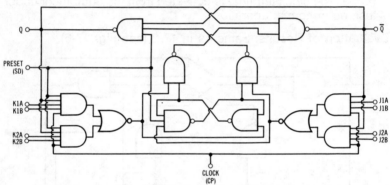

(B) Logic diagram. *(Courtesy Fairchild Camera and Instrument Corp.)*

t_n		t_{n+1}
J	K	Q
0	0	Q_n
0	1	0
1	0	1
1	1	$\overline{Q_n}$

(C) Truth table.

NOTE:
$J = (J1A \cdot J1B) + (J2A \cdot J2B)$
$K = (K1A \cdot K1B) + (K2A \cdot K2B)$
t_n = Bit time before clock pulse
t_{n+1} = Bit time after clock pulse

Positive logic:
LOW input to preset sets Q to HIGH level
Preset is independent of clock

Fig. 17-3. JK master/slave flip-flop with AND-OR inputs.

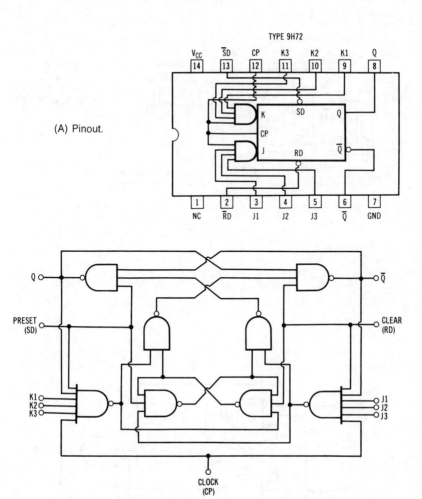

TYPE 9H72

(A) Pinout.

(B) Logic diagram. *(Courtesy Fairchild Camera and Instrument Corp.)*

t_n		t_{n+1}
J	**K**	**Q**
0	0	Q_n
0	1	0
1	0	1
1	1	$\overline{Q}_n$

NOTES:
$J = J1 \cdot J2 \cdot J3$
$K = K1 \cdot K2 \cdot K3$
t_n = Bit time before clock pulse
t_{n+1} = Bit time after clock pulse

Positive logic:
LOW input to preset sets Q to HIGH level
LOW input to clear sets Q to LOW level
Preset and clear are independent of clock

(C) Truth table.

Fig. 17-4. JK master/slave flip-flop with AND inputs.

209

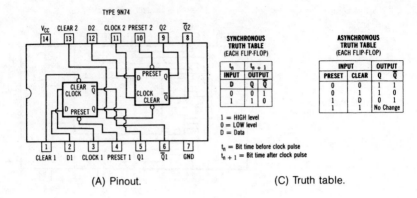

TYPE 9N74

(A) Pinout.   (C) Truth table.

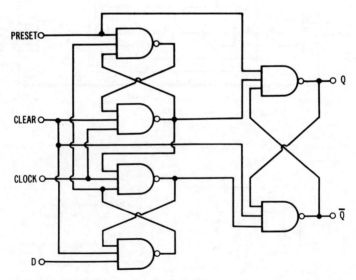

(B) Logic diagram. *(Courtesy Fairchild Camera and Instrument Corp.)*

Fig. 17-5. Dual type-D edge-triggered flip-flop.

210

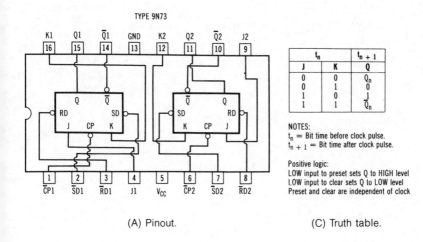

TYPE 9N73

(A) Pinout.

t_n		t_{n+1}
J	K	Q
0	0	Q_n
0	1	0
1	0	1
1	1	$\overline{Q_n}$

NOTES:
t_n = Bit time before clock pulse.
t_{n+1} = Bit time after clock pulse.

Positive logic:
LOW input to preset sets Q to HIGH level
LOW input to clear sets Q to LOW level
Preset and clear are independent of clock

(C) Truth table.

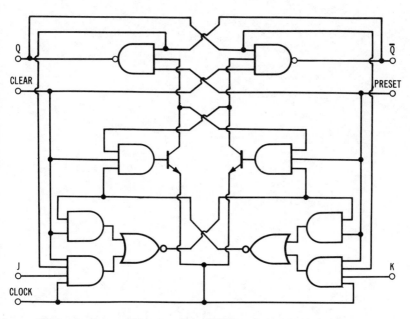

(B) Logic diagram. *(Courtesy Fairchild Camera and Instrument Corp.)*

**Fig. 17-6. JK master/slave flip-flop with
separate presets, clears, and clocks.**

211

Section 18

SHIFT REGISTERS

18-1 To Check the Operation of a 5-Bit Shift Register

Equipment: Logic pulser and logic probe.

Connections Required: Connect V_{cc} and gnd leads of pulser and probe to power supply in the equipment under test. In the first test, the parallel load (P_L) terminal should be tied to a logic low point, and the clear ($\bar{C}_L$) terminal should be tied to a logic high point. In the second test, both the P_L terminal and the $\bar{C}_L$ terminal should be tied to a logic high point.

Procedure: In the first test, inject test pulses at the serial-input (D_S) terminal, and check for pulse output with probe at Q_E terminal. In the second test, inject test pulses successively at input terminals P_A through P_E, and check for output with probe at associated output terminals Q_A through Q_E. The clock input (C_P) is activated from the equipment, as in normal operation.

Evaluation of Results: In normal operation, serial data will be clocked out at the Q_E terminal after a delay of five clock pulses in the first test. Parallel data will be available at a flip-flop output terminal after a delay of one clock pulse. Note that after the register has been parallel loaded, and the $\bar{C}_L$ terminal remains high, but the P_L terminal is changed to a low logic point, the data will normally unload in serial from Q_E after five clock cycles (see Fig. 18-1).

18-2 To Check the Operation of an 8-Bit Shift Register

Equipment: Logic pulser and logic probe.

Connections Required: Connect V_{CC} and gnd leads of probe and pulser to power supply in equipment under test. Tie inputs A and B together.

Procedure: Pulse the tied inputs with the logic pulser, and check for output pulses at Q and $\overline{Q}$. The clock-pulse input ($\overline{CP}$) is activated from the equipment, as in normal operation.

Evaluation of Results: In normal operation, pulses clocked in at the A and B inputs are clocked out at the Q and $\overline{Q}$ outputs after a delay of eight clock pulses (see Fig. 18-2).

18-3 To Check the Operation of an 8-Bit Bidirectional Shift Register

Equipment: Logic pulser and logic probe.

Connections Required: Connect V_{CC} and gnd leads of probe and pulser to power supply in the equipment under test. In the first test, input S_1 should be tied to a low source, and S_0 to a high source. In the second test, these high and low connections are reversed. In the third test, both S_1 and S_0 are tied to a high source.

Procedure: In the first test, apply the pulser at the R input, and apply the probe in turn to each of the outputs. In the second

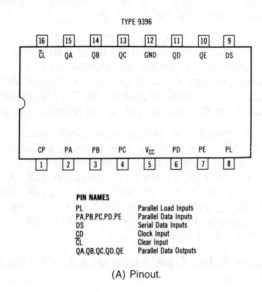

TYPE 9396

(A) Pinout.

Fig. 18-1. A 5-bit

test, apply the pulser at the L input, and apply the probe in turn to each of the outputs. In the third test, apply the pulser in turn to each of the parallel data inputs, while checking at the corresponding output with the probe. The clock input is activated from the equipment, as in normal operation.

Evaluation of Results: In normal operation, serial data will be clocked out at Q_H after a delay of eight clock pulses in the first test. Serial data will be clocked out at the Q_A output after a delay of eight clock pulses in the second test. Parallel data will be unloaded in parallel on the second clock cycle in the third

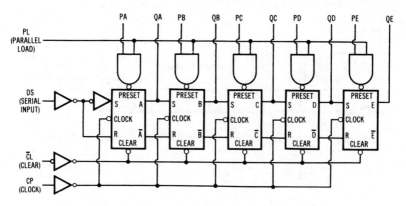

(B) Logic diagram. *(Courtesy Fairchild Camera and Instrument Corp.)*

PRESET COMMON	BIT	CLEAR	SERIAL INPUT	CLOCK	OUTPUT	
0	X	0	X	X	0	Clear all output to logical "1."
1	1	1	X	X	1	Preset outputs to 1 input
1	0	1	X	X	0	bit configuration.
0	X	1	1	enable	1	Serial input shift right.
0	X	1	0	enable	0	Serial-to-parallel conversion.

NOTES:
(a) After loading data, set clear to "1" and preset to "0" clock to give parallel to serial conversion.
(b) Information transferred on rising edge of clock pulse.

(c) Do not enable preset and clears simultaneously. Preset–"1" Clear–"0" = undefined output. Dependent upon which enable is removed first.
(d) X Either logical "0" or logical "1".

(C) Truth table.

shift register.

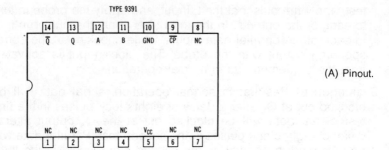

(A) Pinout.

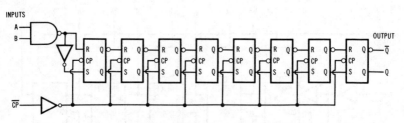

(B) Logic diagram. *(Courtesy Fairchild Camera and Instrument Corp.)*

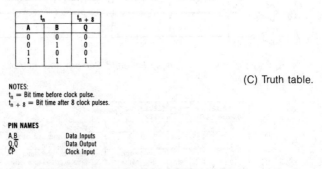

(C) Truth table.

NOTES:
t_n = Bit time before clock pulse.
t_{n+8} = Bit time after 8 clock pulses.

PIN NAMES

A,B	Data Inputs
Q,$\overline{Q}$	Data Output
CP	Clock Input

Fig. 18-2. An 8-bit shift register.

test. If both S_1 and S_0 are held low, parallel-loaded data will be retained until the clock is started again for shift-right, shift-left, or parallel operation (see Fig. 18-3).

18-4 To Check the Operation of a Shift Register With a Logic Comparator

Equipment: Logic comparator as described in the following procedure.

Connections Required: None.

216

Procedure: Plug a known good reference IC into the logic comparator, and place test cable termination over shift-register IC in equipment under test. Turn the equipment on for usual operation.

Evaluation of Results: Any discrepancy in logic states between the reference IC and the IC under test is displayed as an error by a LED, and the corresponding pin is identified in the comparator readout. *Make certain that the comparator contacts are making good contact with the IC pins; otherwise, a false error indication could be obtained.*

24	23	22	21	20	19	18	17	16	15	14	13
V_{CC}	S1	L	PH	QH	PG	QG	PF	QF	PE	QE	$\overline{CL}$

S0	R	PA	QA	PB	QB	PC	QC	PD	QD	CP	GND
1	2	3	4	5	6	7	8	9	10	11	12

PIN NAMES

PA to PH	Parallel Data Inputs
S0,S1	Mode Control Inputs
L	Shift Left Serial Input
R	Shift Right Serial Input
CP	Clock (Active LOW) Input
$\overline{CL}$	Clear (Active LOW) Input
QA to QH	Data Outputs

(A) Pinout.

INPUTS		MODE
S1	S0	
0	0	INHIBIT CLOCK
0	1	SHIFT RIGHT
1	0	SHIFT LEFT
1	1	PARALLEL LOAD

(C) Truth table.

Fig. 18-3. An 8-bit bidirectional shift register.

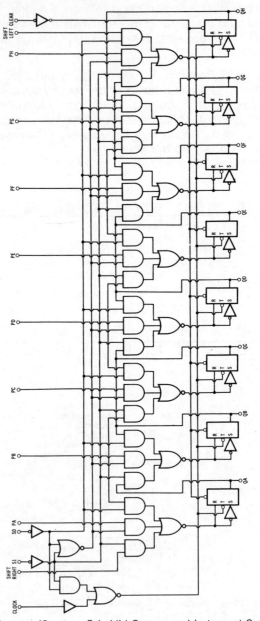

(B) Logic diagram. *(Courtesy Fairchild Camera and Instrument Corp.)*

Fig. 18.3. An 8-bit bidirectional shift register. (Continued)

18-5 To Check the Operation of a Shift Register With a Logic Clip

Equipment: Logic pulser and logic clip.

Connections Required: Same as in 18-1, 2, and 3, except that a probe is not used.

Procedure: Place the logic clip over the IC with the shift register. Stop the circuit under test, and verify the truth table by pulsing the various inputs with the logic pulser.

Evaluation of Results: The state of each pin (high or low) on the IC is shown by the LEDs on top of the logic clip. Unless the responses to the test pulses are in accordance with the truth table, the shift register is defective and should be replaced.

Section 19
COUNTERS

19-1 To Check the Operation of a 4-Bit Binary Counter

Equipment: Logic pulser and logic probe.

Connections Required: The master reset input (MR_1, MR_2) should be tied to a logic LOW point during the test (see Fig. 19-1B). As a 4-bit counter, terminal Q_0 will be externally connected to input $\overline{CP}_1$. Connect V_{cc} and gnd leads of the pulser and probe to the power supply in the equipment under test (see Fig. 19-1).

Procedure: Apply pulser to terminal $\overline{CP}_0$, and inject test pulses. Apply logic probe at output terminal Q_3 to check for counter output. The counter is clocked from the equipment under test, as in normal operation.

Evaluation of Results: In normal operation, an output will be obtained after a delay of eight pulses.

19-2 To Check the Operation of a Decade Counter

Equipment: Logic pulser and logic probe.

Connections Required: With reference to Fig. 19-2, terminals $R_{0(1)}$ and $R_{9(1)}$ should be tied to a low logic point during the test. Connect V_{cc} and gnd leads of the pulser and probe to the power supply in the equipment under test. Note in passing that a symmetrical divide by 10 count is provided when the Q_D output is externally connected to the $\overline{CP}_A$ input.

Procedure: Apply pulser at terminal $\overline{CP}_{BD}$ and inject a train of test pulses. Apply the logic probe at terminal Q_A to check for counter output. The counter is clocked from the equipment under test, as in normal operation.

Evaluation of Results: In normal operation, an output pulse will be obtained for every tenth input pulse.

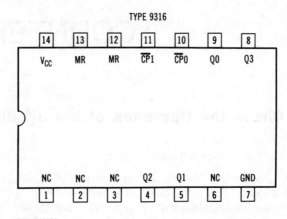

TYPE 9316

PIN NAMES

$\overline{CP0}$	Clock First Stage Negative Edge Input
$\overline{CP1}$	Clock Second, Third, and Fourth Stage Negative Edge Input
MR	"AND" Master Reset to Binary Zero (Asynchronous) Input
Q0,Q1,Q2,Q3	Counter Outputs

(A) Pinout.

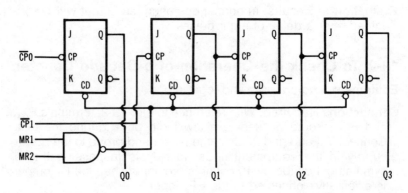

(B) Logic diagram. *(Courtesy Fairchild Camera and Instrument Corp.)*

Fig. 19-1. A 4-bit

19-3 To Check the Operation of a Divide-by-12 Counter

Equipment: Logic pulser and logic probe.

Connections Required: Connect V_{cc} and gnd leads of pulser and probe to the power supply in the equipment under test. The reset input (R_0) should be held low during the test, and may be tied to a fixed logic-low point, if necessary. Note in passing that a divide-by-12 output is provided when the output terminal Q_A is externally connected to input $\overline{CP}_{BC}$ (Fig. 19-3).

Procedure: Apply the pulser at input $\overline{CP}_A$, and inject a train of test pulses. Apply the logic probe at terminal Q_D to check for counter output. The counter is clocked from the equipment under test, as in normal operation.

Evaluation of Results: In normal operation, an output pulse will be obtained for every twelfth input pulse.

(C) Truth table.

COUNT	OUTPUT			
	Q0	Q1	Q2	Q3
0	0	0	0	0
1	1	0	0	0
2	0	1	0	0
3	1	1	0	0
4	0	0	1	0
5	1	0	1	0
6	0	1	1	0
7	1	1	1	0
8	0	0	0	1
9	1	0	0	1
10	0	1	0	1
11	1	1	0	1
12	0	0	1	1
13	1	0	1	1
14	0	1	1	1
15	1	1	1	1

NOTE: Output Q0 connected to input $\overline{CP}$1

(D) Mode selection.

RESET INPUTS		OUTPUTS			
MR1	MR2	Q0	Q1	Q2	Q3
1	1	0	0	0	0
0	1	Count			
1	0	Count			
0	0	Count			

1 = HIGH Voltage Level
0 = LOW Voltage Level
X = Don't Care Condition

binary counter.

19-4 To Check the Operation of a Decade Counter With Strobe Input

Equipment: Logic pulser and logic probe.

Connections Required: With reference to Fig. 19-4, the data strobe and reset inputs should be tied to a low logic point during the test. Connect V_{cc} and gnd leads of the pulser and probe to the power supply in the equipment under test. Note in passing that decade operation is obtained when the Q_D output terminal is externally connected to the clock 1 input.

Procedure: Apply the pulser at clock 2 input and inject a train of test pulses. Apply the logic probe at terminal Q_A to check for counter output. The counter is clocked from the equipment under test, as in normal operation.

Evaluation of Results: In normal operation, an output pulse will be obtained for every tenth input pulse.

19-5 To Check the Operation of a Counter With a Logic Clip

Equipment: Logic clip and logic pulser.

Connections Required: Connect the leads from the logic pulser to the power supply in the equipment under test.

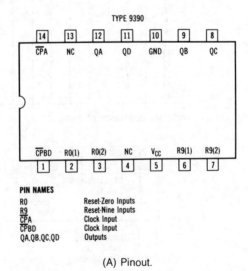

TYPE 9390

| 14 | 13 | 12 | 11 | 10 | 9 | 8 |
| CPA | NC | QA | QD | GND | QB | QC |

| CPBD | R0(1) | R0(2) | NC | V_CC | R9(1) | R9(2) |
| 1 | 2 | 3 | 4 | 5 | 6 | 7 |

PIN NAMES

R0	Reset-Zero Inputs
R9	Reset-Nine Inputs
CPA	Clock Input
CPBD	Clock Input
QA,QB,QC,QD	Outputs

(A) Pinout.

Fig. 19-2. A

Procedure: Stop the clock oscillator in the equipment under test. Use the logic pulser to inject pulses into the clock line; after a pulse is injected, observe the LED readout on the logic clip.

Evaluation of Results: As the counter is being slowly stepped through its entire operating cycle, its outputs, resets, clears, or other operating signals will normally be verified by the truth table.

19-6 To Check the Operation of a Counter With a Logic Comparator

Equipment: Logic comparator.

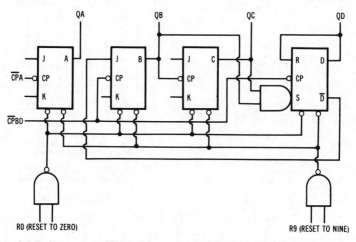

(B) Logic diagram. *(Courtesy Fairchild Camera and Instrument Corp.)*

BCD COUNT SEQUENCE (Note 1)

COUNT	OUTPUT			
	QD	QC	QB	QA
0	0	0	0	0
1	0	0	0	1
2	0	0	1	0
3	0	0	1	1
4	0	1	0	0
5	0	1	0	1
6	0	1	1	0
7	0	1	1	1
8	1	0	0	0
9	1	0	0	1

RESET/COUNT (See Note 2)

RESET INPUTS				OUTPUT			
R0(1)	R0(2)	R9(1)	R9(2)	QD	QC	QB	QA
1	1	0	X	0	0	0	0
1	1	X	0	0	0	0	0
X	X	1	1	1	0	0	1
X	0	X	0	COUNT			
0	X	0	X	COUNT			
0	X	X	0	COUNT			
X	0	0	X	COUNT			

NOTES:
1. Output QA connected to input CPBD for BCD count.
2. X indicates that either a HIGH level or a LOW level may be present.

(C) Truth table.

decade counter.

Connections Required: None.

Procedure: Plug in a known good reference IC into the logic comparator, and place the test cable termination over the counter IC in the equipment under test. Turn the equipment on for usual operation.

Evaluation of Results: Any discrepancy in logic states between the reference IC and the IC under test is displayed as an error by a LED, and the corresponding pin is identified in the comparator readout.

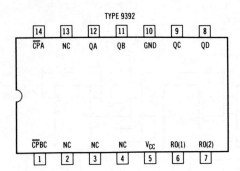

TYPE 9392

14	13	12	11	10	9	8
$\overline{CPA}$	NC	QA	QB	GND	QC	QD

$\overline{CPBC}$	NC	NC	NC	V_{CC}	RO(1)	RO(2)
1	2	3	4	5	6	7

PIN NAMES

RO	Reset-Zero Inputs
$\overline{CPA}$	Clock Input
$\overline{CPBC}$	Clock Input
QA,QB,QC,QD	Count Outputs

(A) Pinout.

COUNT	OUTPUT			
	QD	QC	QB	QA
0	0	0	0	0
1	0	0	0	1
2	0	0	1	0
3	0	0	1	1
4	0	1	0	0
5	0	1	0	1
6	1	0	0	0
7	1	0	0	1
8	1	0	1	0
9	1	0	1	1
10	1	1	0	0
11	1	1	0	1

NOTES:
1. Output QA connected to input $\overline{CPBC}$
2. To reset all outputs to LOW level both RO(1) and RO(2) inputs must be at HIGH level state.
3. Either (or both) resets inputs RO(1) and RO(2) must be at a LOW level to count.

(C) Truth table.

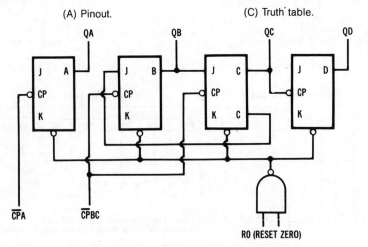

(B) Logic diagram. *(Courtesy Fairchild Camera and Instrument Corp.)*

Fig. 19-3. Typical divide by 12 counter.

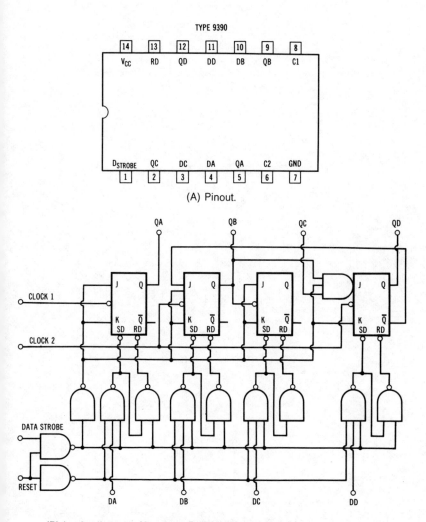

TYPE 9390

(A) Pinout.

(B) Logic diagram. *(Courtesy Fairchild Camera and Instrument Corp.)*

Fig. 19-4. A decade counter with strobe input.

Section 20
DECODERS AND ENCODERS

20-1 To Check the Operation of a 1-of-10 Decoder/ Driver

Equipment: Logic pulser and logic probe.

Connections Required: Connect V_{cc} and gnd leads of the pulser and probe to the power supply in the equipment under test. Tie all four inputs together for the first test (Fig. 20-1). Additional tests are made by tying groups of inputs together as listed in the truth table. Logic low inputs should be tied to a fixed low point during the tests.

Procedure: With all four inputs tied together, inject high and low test pulses from the pulser, and check the outputs with the logic probe. With various inputs tied together in groups, repeat the procedure.

Evaluation of Results: In normal operation, the outputs will correspond to the truth-table listings. Thus, when all four inputs are pulsed high, all outputs are normally high. When all four inputs are pulsed low, all outputs are normally high except $\overline{Q}_0$, which is normally low.

20-2 To Check the Operation of a 1-of-16 Decoder

Equipment: Logic pulser and logic probe.

Connections Required: Connect V_{cc} and gnd leads of the pulser and probe to the power supply in the equipment under test. Tie all four A inputs together for the first test (see Fig. 20-2). The following tests are made by tying groups of inputs together as listed in the truth table. Logic-low inputs should be tied to a fixed low-level point during tests. Inputs $\overline{E}_0$ and $\overline{E}_1$ should be tied to fixed high or low points as indicated in the truth table during the time that the A (address) inputs are pulsed. (An address is a binary number that designates a location where digital information is stored.)

Procedure: With all four A inputs tied together, inject high and low pulses from the logic pulser, and check the outputs with the logic probe. Repeat the procedure with various A inputs tied together.

Evaluation of Results: In normal operation, the outputs will correspond to the truth table listings.

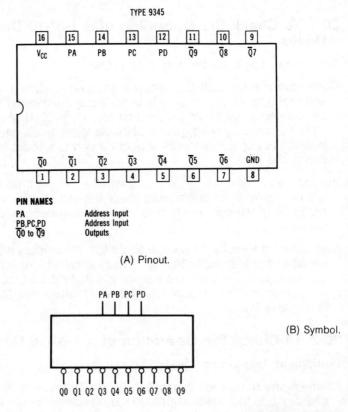

TYPE 9345

16	15	14	13	12	11	10	9
V_{CC}	PA	PB	PC	PD	$\overline{Q}9$	$\overline{Q}8$	$\overline{Q}7$

$\overline{Q}0$	$\overline{Q}1$	$\overline{Q}2$	$\overline{Q}3$	$\overline{Q}4$	$\overline{Q}5$	$\overline{Q}6$	GND
1	2	3	4	5	6	7	8

PIN NAMES

PA   Address Input
PB,PC,PD   Address Input
$\overline{Q}0$ to $\overline{Q}9$ Outputs

(A) Pinout.

PA PB PC PD

Q0 Q1 Q2 Q3 Q4 Q5 Q6 Q7 Q8 Q9

(B) Symbol.

Fig. 20-1. 1-of-10

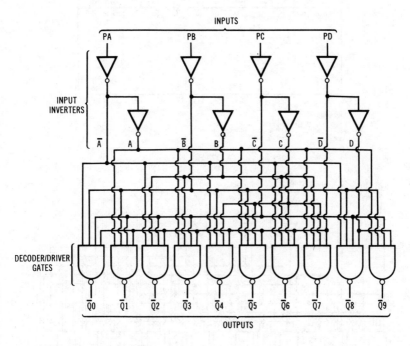

(C) Logic diagram. *(Courtesy Fairchild Camera and Instrument Corp.)*

INPUTS				OUTPUTS									
PD	PC	PB	PA	$\overline{Q0}$	$\overline{Q1}$	$\overline{Q2}$	$\overline{Q3}$	$\overline{Q4}$	$\overline{Q5}$	$\overline{Q6}$	$\overline{Q7}$	$\overline{Q8}$	$\overline{Q9}$
0	0	0	0	0	1	1	1	1	1	1	1	1	1
0	0	0	1	1	0	1	1	1	1	1	1	1	1
0	0	1	0	1	1	0	1	1	1	1	1	1	1
0	0	1	1	1	1	1	0	1	1	1	1	1	1
0	1	0	0	1	1	1	1	0	1	1	1	1	1
0	1	0	1	1	1	1	1	1	0	1	1	1	1
0	1	1	0	1	1	1	1	1	1	0	1	1	1
0	1	1	1	1	1	1	1	1	1	1	0	1	1
1	0	0	0	1	1	1	1	1	1	1	1	0	1
1	0	0	1	1	1	1	1	1	1	1	1	1	0
1	0	1	0	1	1	1	1	1	1	1	1	1	1
1	0	1	1	1	1	1	1	1	1	1	1	1	1
1	1	0	0	1	1	1	1	1	1	1	1	1	1
1	1	0	1	1	1	1	1	1	1	1	1	1	1
1	1	1	0	1	1	1	1	1	1	1	1	1	1
1	1	1	1	1	1	1	1	1	1	1	1	1	1

(D) Truth table.

decoder/driver.

TYPE 9311

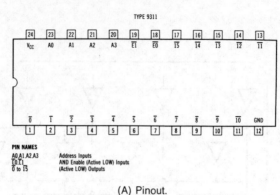

PIN NAMES

A0,A1,A2,A3   Address Inputs
E0,E1   AND Enable (Active LOW) Inputs
0̄ to 1̄5̄   (Active LOW) Outputs

(A) Pinout.

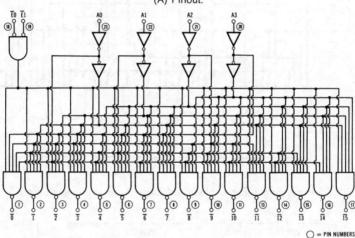

○ = PIN NUMBERS

(B) Logic diagram. *(Courtesy Fairchild Camera and Instrument Corp.)*

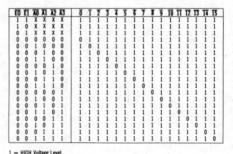

E0	E1	A0	A1	A2	A3	0̄	1̄	2̄	3̄	4̄	5̄	6̄	7̄	8̄	9̄	1̄0̄	1̄1̄	1̄2̄	1̄3̄	1̄4̄	1̄5̄
1	1	X	X	X	X	1	1	1	1	1	1	1	1	1	1	1	1	1	1	1	1
1	0	X	X	X	X	1	1	1	1	1	1	1	1	1	1	1	1	1	1	1	1
0	1	X	X	X	X	1	1	1	1	1	1	1	1	1	1	1	1	1	1	1	1
0	0	0	0	0	0	0	1	1	1	1	1	1	1	1	1	1	1	1	1	1	1
0	0	1	0	0	0	1	0	1	1	1	1	1	1	1	1	1	1	1	1	1	1
0	0	0	1	0	0	1	1	0	1	1	1	1	1	1	1	1	1	1	1	1	1
0	0	1	1	0	0	1	1	1	0	1	1	1	1	1	1	1	1	1	1	1	1
0	0	0	0	1	0	1	1	1	1	0	1	1	1	1	1	1	1	1	1	1	1
0	0	1	0	1	0	1	1	1	1	1	0	1	1	1	1	1	1	1	1	1	1
0	0	0	1	1	0	1	1	1	1	1	1	0	1	1	1	1	1	1	1	1	1
0	0	1	1	1	0	1	1	1	1	1	1	1	0	1	1	1	1	1	1	1	1
0	0	0	0	0	1	1	1	1	1	1	1	1	1	0	1	1	1	1	1	1	1
0	0	1	0	0	1	1	1	1	1	1	1	1	1	1	0	1	1	1	1	1	1
0	0	0	1	0	1	1	1	1	1	1	1	1	1	1	1	0	1	1	1	1	1
0	0	1	1	0	1	1	1	1	1	1	1	1	1	1	1	1	0	1	1	1	1
0	0	0	0	1	1	1	1	1	1	1	1	1	1	1	1	1	1	0	1	1	1
0	0	1	0	1	1	1	1	1	1	1	1	1	1	1	1	1	1	1	0	1	1
0	0	0	1	1	1	1	1	1	1	1	1	1	1	1	1	1	1	1	1	0	1
0	0	1	1	1	1	1	1	1	1	1	1	1	1	1	1	1	1	1	1	1	0

1 = HIGH Voltage Level
0 = LOW Voltage Level
X = Level Does Not Affect Output

(C) Truth table.

Fig. 20-2. 1-of-16 decoder.

20-3 To Check the Operation of a BCD to 7-Segment Decoder

Equipment: Logic pulser and logic probe.

Connections Required: Connect input terminal $\overline{RBI}$ (blanking input), $\overline{BI/RBO}$ (blanking input and ripple-blanking out), and $\overline{LT}$ (lamp test) to a fixed high level during tests. Connect V_{cc} and gnd leads of the pulser and probe to the power supply in the equipment under test. Logic-low inputs should be tied to a fixed low-level point during tests (see Fig. 20-3).

Procedure: With all four inputs tied together, inject high and low pulses from the logic pulser, and check the seven outputs with the logic probe. Repeat the procedure with the various inputs grouped as listed in the truth table.

Evaluation of Results: In normal operation, the outputs will correspond to the truth-table listings.

20-4 To Check the Operation of a Nixie 1-of-10 Decoder/Driver

Equipment: Logic pulser and logic probe.

Connections Required: Connect V_{cc} and gnd leads of pulser and probe to power supply in equipment under test (Fig. 20-4).

Procedure: First, pulse address inputs low; then pulse groups of address inputs high, per truth table, with associated inputs tied to a low logic level. Check corresponding outputs with the logic probe.

Evaluation of Results: In normal operation, the outputs will correspond to the truth-table listings.

20-5 To Check the Operation of a Decoder With a Logic Comparator

Equipment: Logic comparator.

Connections Required: None.

Procedure: Insert a reference decoder IC into the logic comparator, and place the comparator cable termination over the decoder IC under test. Turn the equipment on, and operate it in the usual manner.

Evaluation of Results: Any discrepancy between the high and low states of pins on the IC under test and the reference IC will become apparent as LED indication(s) on the logic comparator. *Follow up as may be required by tests with the logic pulser and probe, pulser and clip, current tracer, or oscilloscope.*

20-6 To Check the Operation of an Eight-Input Priority Encoder

Equipment: Logic pulser and logic probe.

Connections Required: Connect the $\overline{EI}$ (enable input) to a fixed logic-low level. In the first test, all eight inputs are tied together; in the second test, no tie is used; in the third test, input $\overline{6}$ is connected to a fixed logic-low level, and so on. Connect the V_{cc} and gnd leads of the pulser and probe to the power supply in the equipment under test (see Fig. 20-5).

Procedure: In the first test, all eight inputs are pulsed high, and the corresponding output states are checked with the logic probe. In the second test, input $\overline{7}$ is pulsed low, and the output states are checked with the probe. In the third test, the input is held low and input $\overline{7}$ is pulsed high; outputs are checked with the probe, and so on.

Evaluation of Results: In normal operation, the input/output relations will correspond to the listings in the truth table.

20-7 To Check the Operation of an Encoder With a Logic Comparator

Equipment: Logic comparator.

Connections Required: None.

Procedure: Plug in a known good reference IC into the logic comparator and place the test cable termination over the comparator IC in the equipment under test. Turn the equipment on for usual operation.

Evaluation of Results: Any discrepancy in logic states between the reference IC and the IC under test is displayed as an error by a LED, and the corresponding pin is identified in the comparator readout. Since the comparator can only indicate discrepancies in the logic states, other instruments must often be used to pinpoint the cause of the discrepancy. Follow-up tests may be required with the logic probe and pulser, clip and pulser, current tracer, or oscilloscope.

TYPE 9358

(A) Pinout.

DECIMAL OR FUNCTION	$\overline{LT}$	$\overline{RBI}$	D	C	B	A	$\overline{BI/RBO}$	a	b	c	d	e	f	g	NOTE
0	1	1	0	0	0	0	1	1	1	1	1	1	1	0	1
1	1	X	0	0	0	1	1	0	1	1	0	0	0	0	1
2	1	X	0	0	1	0	1	1	1	0	1	1	0	1	
3	1	X	0	0	1	1	1	1	1	1	1	0	0	1	
4	1	X	0	1	0	0	1	0	1	1	0	0	1	1	
5	1	X	0	1	0	1	1	1	0	1	1	0	1	1	
6	1	X	0	1	1	0	1	0	0	1	1	1	1	1	
7	1	X	0	1	1	1	1	1	1	1	0	0	0	0	
8	1	X	1	0	0	0	1	1	1	1	1	1	1	1	
9	1	X	1	0	0	1	1	1	1	1	0	0	1	1	
10	1	X	1	0	1	0	1	0	0	0	1	1	0	1	
11	1	X	1	0	1	1	1	0	0	1	1	0	0	1	
12	1	X	1	1	0	0	1	0	1	0	0	0	1	1	
13	1	X	1	1	0	1	1	1	0	0	1	0	1	1	
14	1	X	1	1	1	0	1	0	0	0	1	1	1	1	
15	1	X	1	1	1	1	1	0	0	0	0	0	0	0	
BI	X	X	X	X	X	X	0	0	0	0	0	0	0	0	2
RBI	1	0	0	0	0	0	0	0	0	0	0	0	0	0	3
LT	0	X	X	X	X	X	1	1	1	1	1	1	1	1	4

NOTES:
1. BI/RBO is wired-AND logic serving as blanking input ($\overline{BI}$) and/or ripple-blanking output ($\overline{RBO}$). The blanking out ($\overline{BI}$) must be open or held at a HIGH level when output functions 0 through 15 are desired, and ripple-blanking input ($\overline{RBI}$) must be open or at a HIGH level if blanking of a decimal 0 is not desired. X = input may be HIGH or LOW.

2. When a LOW level is applied to the blanking input (forced condition) all segment outputs go to a LOW level, regardless of the state of any other input condition.

3. When ripple-blanking input ($\overline{RBI}$) and inputs A,B,C and D are at LOW level, with the lamp test input at HIGH level, all segment outputs go to a HIGH level and the ripple-blanking output ($\overline{RBO}$) goes to a LOW level (response condition).

4. When the blanking input/ripple-blanking output ($\overline{BI/RBO}$) is open or held at a HIGH level, and a LOW level is applied to lamp-test input, all segment outputs go to a LOW level.

(C) Truth table.

Fig. 20-3. A bcd to 7-segment decoder.

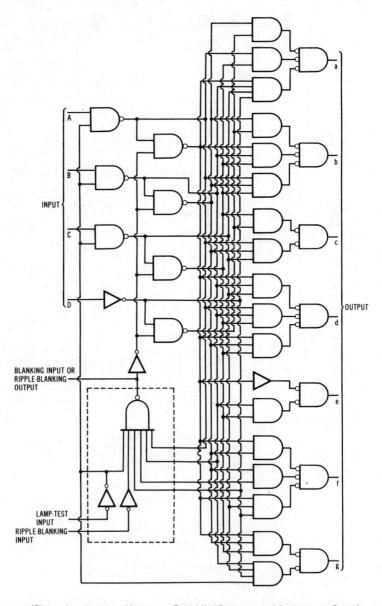

(B) Logic diagram. *(Courtesy Fairchild Camera and Instrument Corp.)*

Fig. 20-3. A bcd to 7-segment decoder. (Continued)

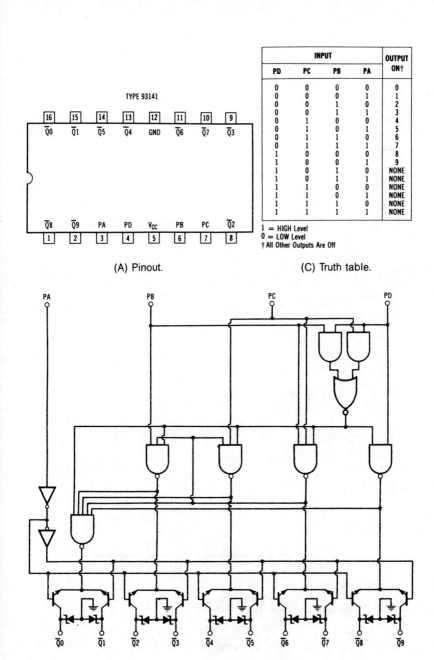

TYPE 93141

(A) Pinout.

	INPUT			OUTPUT
PD	PC	PB	PA	ON†
0	0	0	0	0
0	0	0	1	1
0	0	1	0	2
0	0	1	1	3
0	1	0	0	4
0	1	0	1	5
0	1	1	0	6
0	1	1	1	7
1	0	0	0	8
1	0	0	1	9
1	0	1	0	NONE
1	0	1	1	NONE
1	1	0	0	NONE
1	1	0	1	NONE
1	1	1	0	NONE
1	1	1	1	NONE

1 = HIGH Level
0 = LOW Level
† All Other Outputs Are Off

(C) Truth table.

(B) Logic diagram. *(Courtesy Fairchild Camera and Instrument Corp.)*

Fig. 20-4. A nixie 1-of-10 decoder/driver.

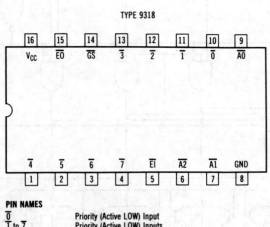

TYPE 9318

16	15	14	13	12	11	10	9
V_{CC}	$\overline{EO}$	$\overline{GS}$	$\overline{3}$	$\overline{2}$	$\overline{1}$	$\overline{0}$	$\overline{A0}$

$\overline{4}$	$\overline{5}$	$\overline{6}$	$\overline{7}$	$\overline{EI}$	$\overline{A2}$	$\overline{A1}$	GND
1	2	3	4	5	6	7	8

PIN NAMES

$\overline{0}$	Priority (Active LOW) Input
$\overline{1}$ to $\overline{7}$	Priority (Active LOW) Inputs
$\overline{EI}$	Enable (Active LOW) Input
$\overline{EO}$	Enable (Active LOW) Output
$\overline{GS}$	Group Select (Active LOW) Output
$\overline{A0},\overline{A1},\overline{A2}$	Address (Active LOW) Outputs

(A) Pinout.

Fig. 20-5. An 8-input

238

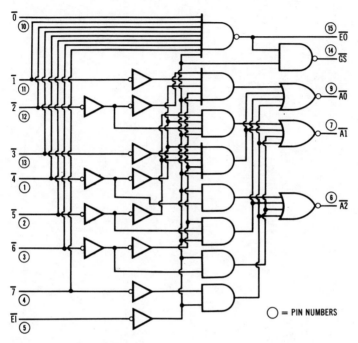

(B) Logic diagram. *(Courtesy Fairchild Camera and Instrument Corp.)*

◯ = PIN NUMBERS

$\overline{EI}$	$\overline{0}$	$\overline{1}$	$\overline{2}$	$\overline{3}$	$\overline{4}$	$\overline{5}$	$\overline{6}$	$\overline{7}$	$\overline{GS}$	$\overline{A0}$	$\overline{A1}$	$\overline{A2}$	$\overline{EO}$
1	X	X	X	X	X	X	X	X	1	1	1	1	1
0	1	1	1	1	1	1	1	1	1	1	1	1	0
0	X	X	X	X	X	X	X	0	0	0	0	0	1
0	X	X	X	X	X	X	0	1	0	1	0	0	1
0	X	X	X	X	X	0	1	1	0	0	1	0	1
0	X	X	X	X	0	1	1	1	0	1	1	0	1
0	X	X	X	0	1	1	1	1	0	0	0	1	1
0	X	X	0	1	1	1	1	1	0	1	0	1	1
0	X	0	1	1	1	1	1	1	0	0	1	1	1
0	0	1	1	1	1	1	1	1	0	1	1	1	1

1 = HIGH Voltage Level
0 = LOW Voltage Level
X = Don't Care

(C) Truth table.

priority encoder.

239

Section 21
MULTIPLEXERS

21-1 To Check the Operation of a Dual Four-Input Multiplexer

Equipment: Logic pulser and logic probe.

Connections Required: In the first test, tie the S_0 and S_1 select inputs to a fixed logic-low point. In the following tests, tie the S_0 and S_1 inputs to logic-high and logic-low points as required by the truth table. Connect the V_{cc} and gnd leads of the pulser and probe to the power supply in the equipment under test (Fig. 21-1).

Procedure: In the first test, use the logic pulser to pulse the I_{oa} input low, and check the logic states at the Z_a and $\overline{Z}_a$ outputs with the logic probe. Continue in a similiar manner until all conditions in the truth table have been checked.

Evaluation of Results: In normal operation, the input/output relations will be in accordance with the truth table listings.

21-2 To Check the Operation of a Quad Two-Input Multiplexer

Equipment: Logic pulser and logic probe.

Connections Required: As specified by the truth table, tie $\overline{E}$ (enable) and S (select) inputs to fixed high or low logic points in consecutive tests. Connect V_{cc} and gnd leads of the pulser and probe to the power supply in equipment under test (Fig. 21-2).

Procedure: Pulse the I_{oc} or I_{ic} inputs high or low with the logic pulser as required by the truth table in consecutive tests. Check the corresponding outputs at the Z_c output with the logic probe.

Evaluation of Results: In normal operation, the input/output relations will be in accordance with the truth-table listings.

21-3 To Check the Operation of an Eight-Input Multiplexer

Equipment: Logic pulser and logic probe.

Connections Required: As specified by the truth table, tie $\bar{E}$, S_2, S_1, and S_0 inputs to fixed high or low logic points in consecutive tests. Connect V_{cc} and gnd leads of the pulser and probe to the power supply in equipment under test (see Fig. 21-3).

Procedure: Pulse the I_0 through I_7 inputs high or low with the logic pulser as required by the truth table in consecutive tests. Check the corresponding outputs at Z and $\bar{Z}$ with the logic probe.

Evaluation of Results: In normal operation, the input/output relations will be in accordance with the truth-table listings.

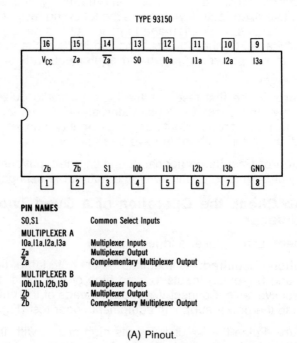

TYPE 93150

16	15	14	13	12	11	10	9
V_{CC}	Za	$\overline{Za}$	S0	I0a	I1a	I2a	I3a

Zb	$\overline{Zb}$	S1	I0b	I1b	I2b	I3b	GND
1	2	3	4	5	6	7	8

PIN NAMES

S0,S1   Common Select Inputs

MULTIPLEXER A
I0a,I1a,I2a,I3a   Multiplexer Inputs
Za   Multiplexer Output
$\overline{Za}$ Complementary Multiplexer Output

MULTIPLEXER B
I0b,I1b,I2b,I3b   Multiplexer Inputs
Zb   Multiplexer Output
$\overline{Zb}$ Complementary Multiplexer Output

(A) Pinout.

Fig. 21-1. A dual

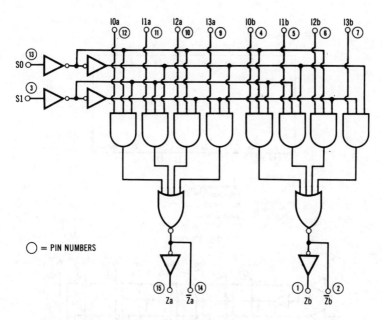

○ = PIN NUMBERS

(B) Logic diagram. *(Courtesy Fairchild Camera and Instrument Corp.)*

SELECTED INPUTS		INPUTS				OUTPUTS	
S0	S1	I0a	I1a	I2a	I3a	Za	$\overline{Za}$
0	0	0	X	X	X	0	1
0	0	1	X	X	X	1	0
1	0	X	0	X	X	0	1
1	0	X	1	X	X	1	0
0	1	X	X	0	X	0	1
0	1	X	X	1	X	1	0
1	1	X	X	X	0	0	1
1	1	X	X	X	1	1	0
S0	S1	I0b	I1b	I2b	I3b	Zb	$\overline{Zb}$
0	0	0	X	X	X	0	1
0	0	1	X	X	X	1	0
1	0	X	0	X	X	0	1
1	0	X	1	X	X	1	0
0	1	X	X	0	X	0	1
0	1	X	X	1	X	1	0
1	1	X	X	X	0	0	1
1	1	X	X	X	1	1	0

1 = HIGH Voltage Level
0 = LOW Voltage Level
X = Either HIGH or LOW Logic Level

(C) Truth table.

4-input multiplexer.

TYPE 9322

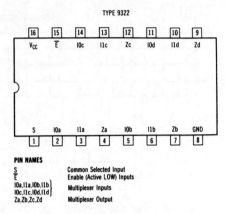

PIN NAMES

S	Common Selected Input
Ē	Enable (Active LOW) Inputs
I0a,I1a,I0b,I1b I0c,I1c,I0d,I1d	Multiplexer Inputs
Za,Zb,Zc,Zd	Multiplexer Output

(A) Pinout.

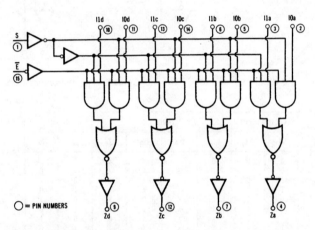

○ = PIN NUMBERS

(B) Logic diagram. *(Courtesy Fairchild Camera and Instrument Corp.)*

ENABLE	SELECT INPUT	INPUTS		OUTPUT
E	S	I0X	I1X	ZX
1	X	X	X	0
0	1	X	0	0
0	1	X	1	1
0	0	0	X	0
0	0	1	X	1

1 = HIGH Voltage Level
0 = LOW Voltage Level
X = Either HIGH or LOW

(C) Truth table.

Fig. 21-2. A quad 2-input multiplexer.

244

21-4 To Check the Operation of a 16-Input Multiplexer

Equipment: Logic pulser and logic probe.

Connections Required: Tie the strobe ($\overline{S}$), A, B, C, and D inputs to fixed high or low logic points in consecutive tests as noted by the truth table. Connect V_{cc} and gnd leads of the pulser and probe to the power supply in equipment under test (see Fig. 21-4).

Procedure: Inject test pulses with the logic pulser at inputs E_0 through E_{15}, as specified by the truth table in consecutive tests. Use the logic probe to check the corresponding outputs at the W output terminal.

Evaluation of Results: In normal operation, the input/output relations will be in accordance with the truth-table listings.

21-5 To Check the Operation of a Multiplexer With a Logic Comparator

Equipment: Logic Comparator.

Connections Required: None.

Procedure: Plug a known good reference IC into the logic comparator and place the test-cable termination over the multiplexer IC in equipment under test. Turn the equipment on for usual operation.

Evaluation of Results: Any discrepancy in logic states between the reference IC and the IC under test is displayed as an error by an LED, and the corresponding pin is identified in the comparator readout. An open bond internal to the IC or a solder bridge external to the IC are typical causes of a discrepancy in logic states. However, a logic comparator cannot distinguish between these, or other, faults internal or external to the IC under test. Therefore, the technician must usually make signal-injection and signal-tracing tests in the circuitry associated with the incorrect logic level(s).

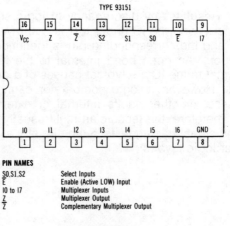

TYPE 93151

| 16 | 15 | 14 | 13 | 12 | 11 | 10 | 9 |
| Vcc | Z | Z̄ | S2 | S1 | S0 | Ē | I7 |

| I0 | I1 | I2 | I3 | I4 | I5 | I6 | GND |
| 1 | 2 | 3 | 4 | 5 | 6 | 7 | 8 |

PIN NAMES

S0,S1,S2	Select Inputs
Ē	Enable (Active LOW) Input
I0 to I7	Multiplexer Inputs
Z	Multiplexer Output
Z̄	Complementary Multiplexer Output

(A) Pinout.

Fig. 21-3. An 8-input

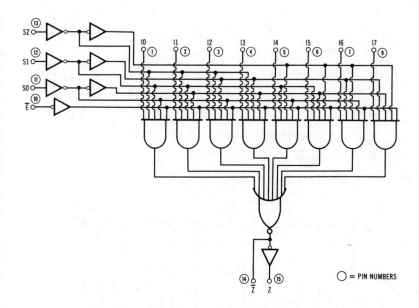

(B) Logic diagram. *(Courtesy Fairchild Camera and Instrument Corp.)*

(C) Truth table.

Ē	S2	S1	S0	I0	I1	I2	I3	I4	I5	I6	I7	Z̄	Z
1	X	X	X	X	X	X	X	X	X	X	X	1	0
0	0	0	0	0	X	X	X	X	X	X	X	1	0
0	0	0	0	1	X	X	X	X	X	X	X	0	1
0	0	0	1	X	0	X	X	X	X	X	X	1	0
0	0	0	1	X	1	X	X	X	X	X	X	0	1
0	0	1	0	X	X	0	X	X	X	X	X	1	0
0	0	1	0	X	X	1	X	X	X	X	X	0	1
0	0	1	1	X	X	X	0	X	X	X	X	1	0
0	0	1	1	X	X	X	1	X	X	X	X	0	1
0	1	0	0	X	X	X	X	0	X	X	X	1	0
0	1	0	0	X	X	X	X	1	X	X	X	0	1
0	1	0	1	X	X	X	X	X	0	X	X	1	0
0	1	0	1	X	X	X	X	X	1	X	X	0	1
0	1	1	0	X	X	X	X	X	X	0	X	1	0
0	1	1	0	X	X	X	X	X	X	1	X	0	1
0	1	1	1	X	X	X	X	X	X	X	0	1	0
0	1	1	1	X	X	X	X	X	X	X	1	0	1

1 = HIGH Voltage Level
0 = LOW Voltage Level
X = Level Does Not Affect Output

multiplexer.

TYPE 74150

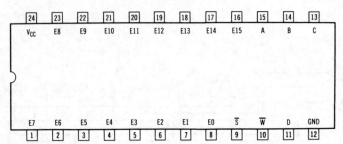

PIN NAMES

E0 to E15	Data Inputs
$\overline{S}$	Strobe (Enable) Input
A,B,C,D	Data Select Inputs
$\overline{W}$	Data Output
Y	Data Output

(A) Pinout.

					INPUTS																OUTPUT
D	C	B	A	STROBE	E0	E1	E2	E3	E4	E5	E6	E7	E8	E9	E10	E11	E12	E13	E14	E15	W
X	X	X	X	1	X	X	X	X	X	X	X	X	X	X	X	X	X	X	X	X	1
0	0	0	0	0	0	X	X	X	X	X	X	X	X	X	X	X	X	X	X	X	1
0	0	0	0	0	1	X	X	X	X	X	X	X	X	X	X	X	X	X	X	X	0
0	0	0	1	0	X	0	X	X	X	X	X	X	X	X	X	X	X	X	X	X	1
0	0	0	1	0	X	1	X	X	X	X	X	X	X	X	X	X	X	X	X	X	0
0	0	1	0	0	X	X	0	X	X	X	X	X	X	X	X	X	X	X	X	X	1
0	0	1	0	0	X	X	1	X	X	X	X	X	X	X	X	X	X	X	X	X	0
0	0	1	1	0	X	X	X	0	X	X	X	X	X	X	X	X	X	X	X	X	1
0	0	1	1	0	X	X	X	1	X	X	X	X	X	X	X	X	X	X	X	X	0
0	1	0	0	0	X	X	X	X	0	X	X	X	X	X	X	X	X	X	X	X	1
0	1	0	0	0	X	X	X	X	1	X	X	X	X	X	X	X	X	X	X	X	0
0	1	0	1	0	X	X	X	X	X	0	X	X	X	X	X	X	X	X	X	X	1
0	1	0	1	0	X	X	X	X	X	1	X	X	X	X	X	X	X	X	X	X	0
0	1	1	0	0	X	X	X	X	X	X	0	X	X	X	X	X	X	X	X	X	1
0	1	1	0	0	X	X	X	X	X	X	1	X	X	X	X	X	X	X	X	X	0
0	1	1	1	0	X	X	X	X	X	X	X	0	X	X	X	X	X	X	X	X	1
0	1	1	1	0	X	X	X	X	X	X	X	1	X	X	X	X	X	X	X	X	0
1	0	0	0	0	X	X	X	X	X	X	X	X	0	X	X	X	X	X	X	X	1
1	0	0	0	0	X	X	X	X	X	X	X	X	1	X	X	X	X	X	X	X	0
1	0	0	1	0	X	X	X	X	X	X	X	X	X	0	X	X	X	X	X	X	1
1	0	0	1	0	X	X	X	X	X	X	X	X	X	1	X	X	X	X	X	X	0
1	0	1	0	0	X	X	X	X	X	X	X	X	X	X	0	X	X	X	X	X	1
1	0	1	0	0	X	X	X	X	X	X	X	X	X	X	1	X	X	X	X	X	0
1	0	1	1	0	X	X	X	X	X	X	X	X	X	X	X	0	X	X	X	X	1
1	0	1	1	0	X	X	X	X	X	X	X	X	X	X	X	1	X	X	X	X	0
1	1	0	0	0	X	X	X	X	X	X	X	X	X	X	X	X	0	X	X	X	1
1	1	0	0	0	X	X	X	X	X	X	X	X	X	X	X	X	1	X	X	X	0
1	1	0	1	0	X	X	X	X	X	X	X	X	X	X	X	X	X	0	X	X	1
1	1	0	1	0	X	X	X	X	X	X	X	X	X	X	X	X	X	1	X	X	0
1	1	1	0	0	X	X	X	X	X	X	X	X	X	X	X	X	X	X	0	X	1
1	1	1	0	0	X	X	X	X	X	X	X	X	X	X	X	X	X	X	1	X	0
1	1	1	1	0	X	X	X	X	X	X	X	X	X	X	X	X	X	X	X	0	1
1	1	1	1	0	X	X	X	X	X	X	X	X	X	X	X	X	X	X	X	1	0

When used to indicate an input condition, X = LOGICAL 1 or LOGICAL 0.

(C) Truth table.

Fig. 21-4. A 16-input

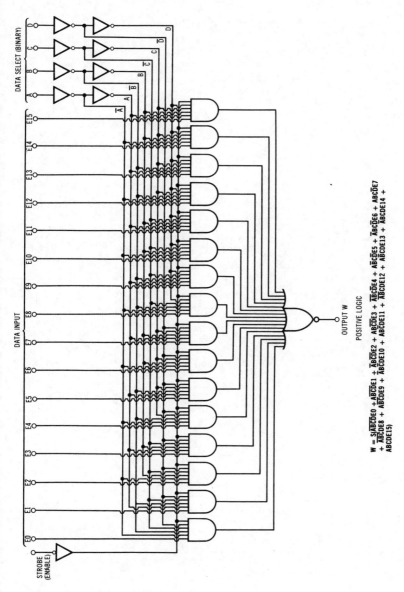

$$W = S(A\overline{B}\overline{C}\overline{D}E0 + A\overline{B}\overline{C}\overline{D}E1 + \overline{A}B\overline{C}\overline{D}E2 + AB\overline{C}\overline{D}E3 + \overline{A}\overline{B}C\overline{D}E4 + A\overline{B}C\overline{D}E5 + \overline{A}BC\overline{D}E6 + ABC\overline{D}E7 + \overline{A}\overline{B}\overline{C}DE8 + A\overline{B}\overline{C}DE9 + \overline{A}B\overline{C}DE10 + AB\overline{C}DE11 + \overline{A}\overline{B}CDE12 + A\overline{B}CDE13 + \overline{A}BCDE14 + ABCDE15)$$

POSITIVE LOGIC

(B) Logic diagram. *(Courtesy Fairchild Camera and Instrument Corp.)*

multiplexer.

249

Section 22

COMPARATORS

22-1 To Check the Operation of a 4-Bit Quad Exclusive-NOR Comparator

Equipment: Logic pulser and logic probe.

Connections Required: Connect V_{cc} and gnd leads of the pulser and probe to the power supply in the equipment under test. In the first test, tie the two inputs together. In the second test, tie one of the inputs to a fixed low level point. In the third test, tie the other input to a fixed low level point (Fig. 22-1).

Procedure: Pulse the inputs high consecutively with the logic pulser, in accordance with the truth table. Check the corresponding outputs with the logic probe.

Evaluation of Results: In normal operation, the input/output relations will be in accordance with the truth table.

22-2 To Check the Operation of a 5-Bit Comparator

Equipment: Logic comparator.

Connections Required: None.

Procedure: Plug a known good reference IC into the logic comparator and place the test-cable termination over the comparator IC in the equipment under test. Turn the equipment on for usual operation (see Fig. 22-2).

Evaluation of Results: Any discrepancy in logic states between the reference IC and the IC under test is displayed as an error by an LED, and the corresponding pin is identified in the comparator readout. Follow up with logic probe, pulser, tracer, and scope tests as required.

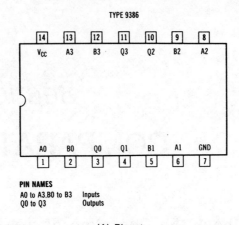

TYPE 9386

PIN NAMES
A0 to A3,B0 to B3   Inputs
Q0 to Q3   Outputs

(A) Pinout.

(B) Logic diagram. *(Courtesy Fairchild Camera and Instrument Corp.)*

INPUTS		OUTPUT
A	B	Q
0	0	1
1	0	0
0	1	0
1	1	1

(C) Truth table.

Fig. 22-1. A 4-bit quad exclusive-NOR comparator.

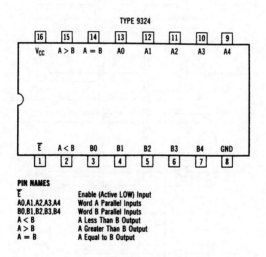

TYPE 9324

16	15	14	13	12	11	10	9
V_{CC}	A > B	A = B	A0	A1	A2	A3	A4

$\overline{E}$	A < B	B0	B1	B2	B3	B4	GND
1	2	3	4	5	6	7	8

PIN NAMES

$\overline{E}$	Enable (Active LOW) Input
A0,A1,A2,A3,A4	Word A Parallel Inputs
B0,B1,B2,B3,B4	Word B Parallel Inputs
A < B	A Less Than B Output
A > B	A Greater Than B Output
A = B	A Equal to B Output

(A) Pinout.

(C) Truth table.

E	Ay	By	A < B	A > B	A = B
1	X	X	0	0	0
0	Word A = Word B		0	0	1
0	Word A > Word B		0	1	0
0	Word B > Word A		1	0	0

1 = HIGH Voltage Level
0 = LOW Voltage Level
X = Either HIGH or LOW Voltage Level

Fig. 22-2. A 5-bit comparator.

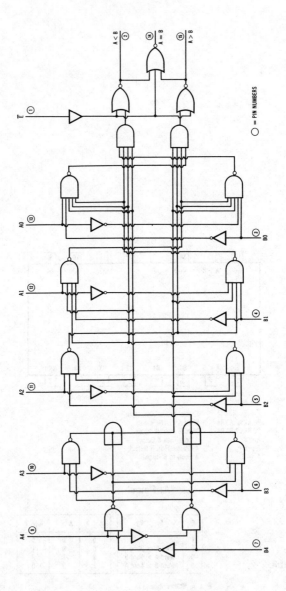

(B) Logic diagram. *(Courtesy Fairchild Camera and Instrument Corp.)*

Fig. 22-2. A 5-bit comparator. (Continued)

Section 23
PARITY GENERATOR/CHECKERS

23-1 To Test an 8-Bit Parity Generator/Checker

Equipment: Logic pulser and logic probe.

Connections Required: With reference to the truth table, tie the odd-input terminal to data inputs 2 through 7, and to a fixed low-logic point. Tie data inputs 0 and 1 together. These ties prepare for the first step; succeeding steps are made in a similar manner. Connect V_{cc} and gnd leads of the pulser and probe to the power supply in the equipment under test (Fig. 23-1).

Procedure: Apply the logic pulser to the tied 0 and 1 data inputs and inject a high pulse. Check the output at the Σ even and the Σ odd terminals with the logic probe.

Evaluation of Results: In normal operation, the input/output relations will be in accordance with the truth table.

23-2 To Test a 9-Bit Parity Generator/Checker

Equipment: Logic pulser and logic probe.

Connections Required: Tie the enable ($\overline{E}$) terminal to a fixed logic-low source; tie the inputs in even and odd groups for consecutive tests (see Fig. 23-2). A group to be tested with low input is tied to a fixed logic-low source. Connect V_{cc} and gnd leads of pulser and probe to power supply in the equipment under test.

Procedure: Apply the logic pulser to the terminal(s) to be pulsed

TYPE 93180

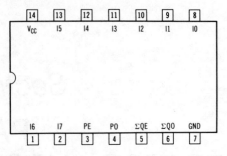

PIN NAMES

I0 to I7	Parity Inputs
PO	Odd Parity Input
PE	Even Parity Input
ΣQ0	Sum Odd Outputs
ΣQE	Sum Even Outputs

(A) Pinout.

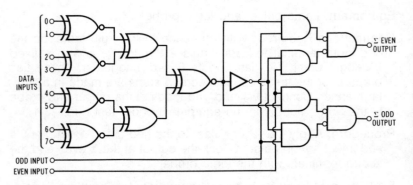

(B) Logic diagram. *(Courtesy Fairchild Camera and Instrument Corp.)*

INPUTS			OUTPUTS	
Σ of 1'S 0 THRU 7	EVEN	ODD	Σ EVEN	Σ ODD
EVEN	1	0	1	0
ODD	1	0	0	1
EVEN	0	1	0	1
ODD	0	1	1	0
X	1	1	0	0
X	0	0	1	1

X = Irrelevant

(C) Truth table.

Fig. 23-1. An 8-bit parity generator/checker.

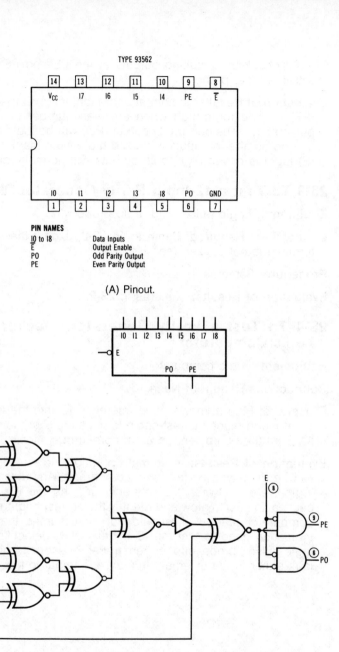

TYPE 93562

PIN NAMES

10 to 18	Data Inputs
E	Output Enable
PO	Odd Parity Output
PE	Even Parity Output

(A) Pinout.

(B) Symbol.

(C) Logic diagram. *(Courtesy Fairchild Camera and Instrument Corp.)*

Fig. 23-2. A 9-bit parity generator/checker.

257

high; check the resulting outputs at the PE and PO terminals with the logic probe.

Evaluation of Results: In normal operation, the even-parity output (PE) will be logic-high when an even number of inputs are pulsed high. The odd-parity output (PO) will be logic high when an odd number of inputs is pulsed high. Note that if the enable (Ē) input is driven high, both outputs will normally go low.

23-3 To Test a 12-Input Parity Generator/Checker

Equipment: Logic pulser and logic probe.

Connections Required: Same as in 23-2, except that an enable terminal is not utilized (see Fig. 23-3).

Procedure: Same as in 23-2.

Evaluation of Results: Same as in 23-2.

23-4 To Test a Parity Generator/Checker With a Logic Comparator

Equipment: Logic comparator.

Connections Required: None.

Procedure: Plug a known good reference IC into the logic comparator and place the test-cable termination over the suspected IC. Turn the equipment on as in normal operation.

Evaluation of Results: In normal operation, no error indications will be displayed by the logic comparator. Any discrepancy in logic states between the reference IC and the IC under test causes an LED to glow and identify the corresponding pin in the comparator readout. It should not be assumed that an error indication denotes that the IC under test is defective. Use the logic pulser, probe, clip, current tracer, or scope to pinpoint the fault either in the external circuitry or internal to the IC.

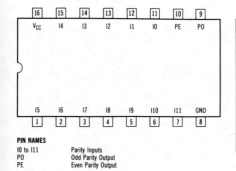

TYPE 9348

PIN NAMES
I0 to I11   Parity Inputs
PO   Odd Parity Output
PE   Even Parity Output

(A) Pinout.

INPUTS		OUTPUTS	
I0,I1,I2,I3,I4,I5,I6,I7,I8,I9,I10,I11		PO	PE
All Twelve	Inputs LOW	0	1
Any One	Input HIGH	1	0
Any Two	Inputs HIGH	0	1
Any Three	Inputs HIGH	1	0
Any Four	Inputs HIGH	0	1
Any Five	Inputs HIGH	1	0
Any Six	Inputs HIGH	0	1
Any Seven	Inputs HIGH	1	0
Any Eight	Inputs HIGH	0	1
Any Nine	Inputs HIGH	1	0
Any Ten	Inputs HIGH	0	1
Any Eleven	Inputs HIGH	1	0
All Twelve	Inputs HIGH	0	1

(C) Truth table.

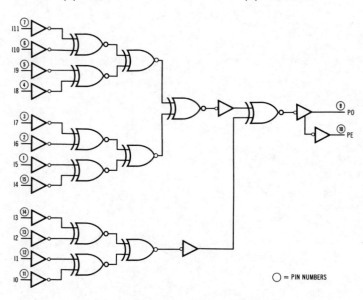

(B) Logic diagram. *(Courtesy Fairchild Camera and Instrument Corp.)*

Fig. 23-3. A 12-bit parity generator/checker.

Section 24

SCHMITT TRIGGERS AND MONOSTABLES

24-1 To Check the Operation of a Dual Four-Input NAND Schmitt Trigger

Equipment: Logic pulser and logic probe.

Connections Required: In the first test,. tie all four inputs of a Schmitt trigger together. In follow-up tests, tie consecutive inputs to a fixed logic-low level. Connect V_{cc} and gnd leads of the pulser and probe to the power supply in the equipment under test (see Fig. 24-1).

Procedure: In the first test, inject positive pulses followed by negative pulses into the four tied inputs. Observe the resulting outputs with the logic probe. Then, repeat the test as consecutive inputs are connected to a low-level point.

Evaluation of Results: In normal operation, an output is obtained only when all four inputs are simultaneously pulsed high, and then low. Each injected pulse of opposite level is normally followed by an output pulse of opposite level.

24-2 To Check the Operation of a Monostable Multivibrator

Equipment: Logic comparator.

Connections Required: None.

Procedure: Place a known good reference IC into the logic comparator and put the test-cable termination over the IC under test. Turn the equipment on as in normal operation (see Fig. 24-2).

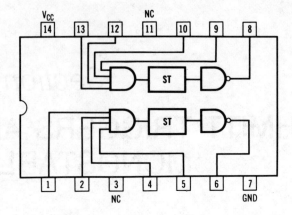

TYPE 9N13

Positive logic: $Y = \overline{ABCD}$
NC — No internal connection

Fig. 24-1. Pinout for a Schmitt trigger.

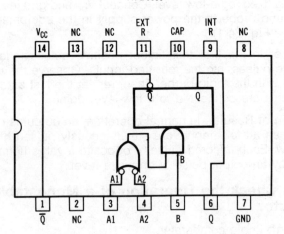

TYPE 9603

NOTE: 9-11 Timing Pins

(A) Pinout.

Fig. 24-2. A mono-

Evaluation of Results: In normal operation, no error indication will be displayed by the logic comparator. Any discrepancy between logic states between the reference IC and the IC under test causes an LED to glow and identify the corresponding pin in the comparator readout. An error indication usually requires follow-up tests with the logic probe and pulser, clip, current tracer, or oscilloscope.

t_n INPUT			t_{n+1} INPUT			OUTPUT
A1	A2	B	A1	A2	B	
1	1	0	1	1	1	Inhibit
0	X	1	0	X	0	Inhibit
X	0	1	X	0	0	Inhibit
0	X	0	0	X	1	One Shot
X	0	0	X	0	1	One Shot
1	1	1	X	0	1	One Shot
1	1	1	0	X	1	One Shot
X	0	0	X	1	0	Inhibit
0	X	0	1	X	0	Inhibit
X	0	1	1	1	1	Inhibit
0	X	1	1	1	1	Inhibit
1	1	0	X	0	0	Inhibit
1	1	0	0	X	0	Inhibit

$1 = V_{IH} \geqslant 2V$
$0 = V_{IL} \leqslant 0.8V$

Positive logic: See truth table and notes 5 and 6.

NOTES:
1. t_n = time before input transition.
2. t_{n+1} = time after input transition.
3. X indicates that either a HIGH or LOW, may be present.
4. NC = No Internal Connection.
5. A1 and A2 are negative edge triggered-logic inputs, and will trigger the one shot when either or both go to LOW level with B at HIGH level.
6. B is positive Schmitt-trigger input for slow edges or level detection and will trigger the one shot when B goes to HIGH level with either A1 or A2 at LOW level. (See Truth Table).
7. External timing capacitor may be connected between pin 10 (positive) and pin 11. With no external capacitance, an output pulse width of typically 30 ns is obtained.
8. To use the internal timing resistor (2kΩ nominal), connect pin 9 to pin 14.
9. To obtain variable pulse width connect external variable resistance between pin 9 and pin 14. No external current limiting is needed.
10. For accurate repeatable pulse widths connect an external resistor between pin 11 and pin 14 with pin 9 open-circuit.

(B) Truth table.

stable multivibrator.

24-3 To Check the Operation of a Retriggerable Monostable Multivibrator

Equipment: Logic comparator.

Connections Required: None.

Procedure: Insert a known good reference IC into the logic comparator and place the test-cable termination over the IC under test. Turn the equipment on as in normal operation (see Fig. 24-3).

Evaluation of Results: No error indication will be displayed by the logic comparator if the IC under test is normal. Any discrepancy between logic states of the reference IC and the IC under

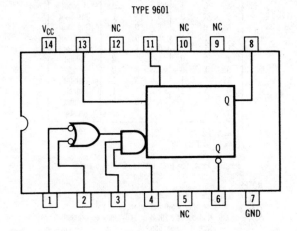

(A) Pinout.

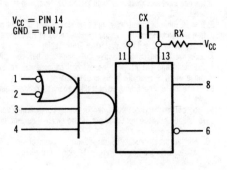

(B) Symbol.

Fig. 24-3. Retriggerable monostable multivibrator.

test will cause an LED to glow and to identify the corresponding pin in the comparator readout. Make certain that a complete test has been run on the retriggerable monostable multivibrator; for example, it might be determined that operation is normal on leading-edge triggering, but is faulty on trailing-edge triggering. Similarly, make certain that the test run includes activation of the retriggering function.

24-4 To Check the Operation of a Retriggerable, Resettable Monostable Multivibrator

Equipment: Logic Comparator.

Connections Required: None.

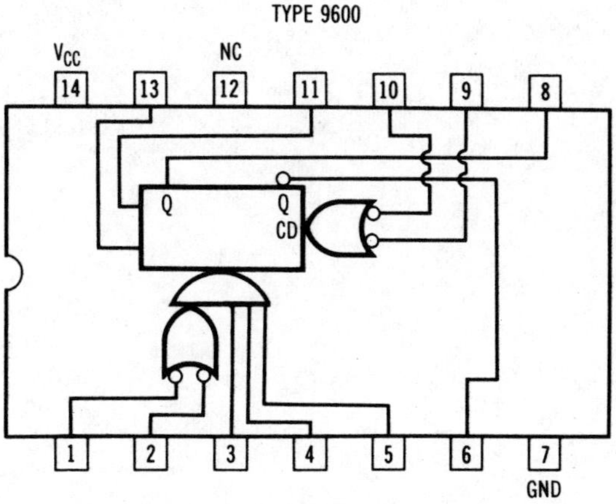

(A) Pinout.

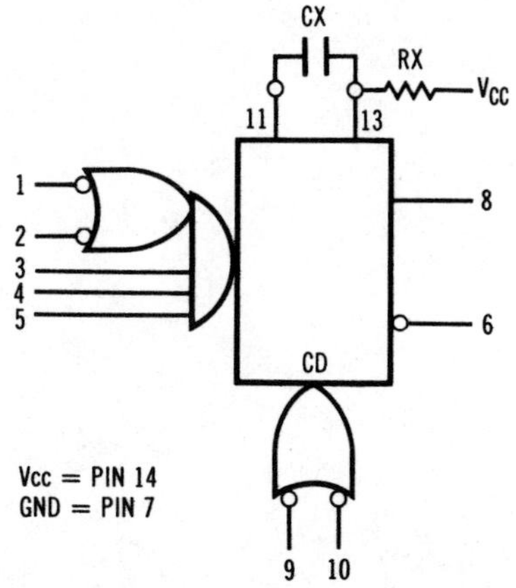

(B) Symbol.

Fig. 24-4. Retriggerable resettable monostable multivibrator.

265

Procedure: Plug a known good reference IC into the logic comparator and push the test-cable termination over the IC under test. Turn the equipment on as for normal operation (see Fig. 24-4).

Evaluation of Results: In normal operation, no error indication will be displayed by the logic comparator. Any discrepancy between logic states of the IC under test and the reference IC will cause an LED to glow and thereby to identify the pin that has an abnormal response. Since an indicated fault may be either internal or external to the IC under test, follow-up tests are often required with the logic pulser and probe, clip, current tracer, or oscilloscope. Also, make a thorough test run—be sure that the data inputs include all of the IC functions.

Section 25
EXPANDABLE GATES

25-1 To Check the Operation of an Expandable Two-Wide, Two-Input AND-OR-INVERT Gate

Equipment: Logic pulser and logic probe.

Connections Required: Connect V_{cc} and gnd lead of the pulser and probe to the power supply in the equipment under test.

Procedure: The same general procedure is followed as previously explained in Section 20. In addition, with the logic probe applied at the Y output, a high output normally results when the X input is pulsed high, and a high output results when the $\overline{X}$ input is pulsed low (see Fig. 25-1).

Evaluation of Results: In normal operation, the expandable gate has input/output relations in accordance with its logic equation.

25-2 To Check the Operation of a Dual Four-Input Expander

Equipment: Logic pulser and logic probe.

Connections Required: Connect V_{cc} and gnd terminals of the pulser and probe to the power supply in the equipment under test. Observe that the X and $\overline{X}$ output terminals of the expander will be connected to the X and $\overline{X}$ input pins of an AND-OR-INVERT gate, as described in 25-1.

Procedure: Same as in 25-1.

Evaluation of Results: In normal operation, the expander will show input/output relations in accordance with its logic equation (see Fig. 25-2).

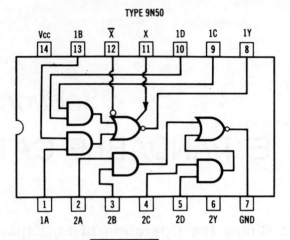

(A) Pinout.

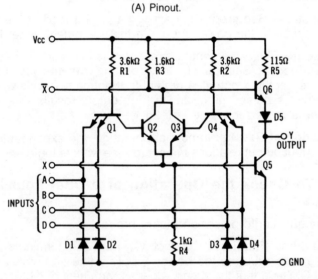

(B) Logic diagram. *(Courtesy Fairchild Camera and Instrument Corp.)*

Fig. 25-1. An expandable 2-wide 2-input AND-OR-INVERT gate.

25-3 To Check the Operation of an Expandable 2-2-2-3-Input AND-OR Gate

Equipment: Logic pulser and logic probe.

TYPE 9N60

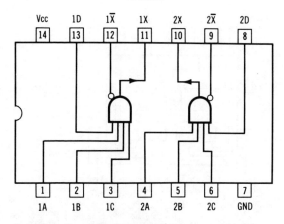

Positive logic: **X = ABCD when connected to X and $\overline{X}$ pins of an expandable gate.**

(A) Pinout.

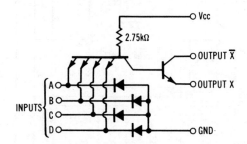

(B) Logic diagram. *(Courtesy Fairchild Camera and Instrument Corp.)*

Fig. 25-2. A dual 4-input expander.

Connections Required: Connect V_{cc} and gnd leads of the pulser and probe to the power supply in the equipment under test (see Fig. 25-3).

Procedure: Follow the same general test procedure as in Section 21-2.

Evaluation of Results: In normal operation, the expandable gate will have input/output relations in accordance with its logic equation.

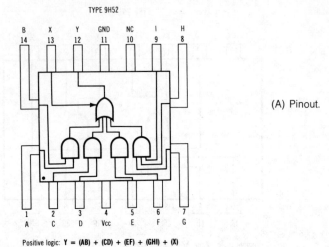

TYPE 9H52

(A) Pinout.

Positive logic: Y = (AB) + (CD) + (EF) + (GHI) + (X)

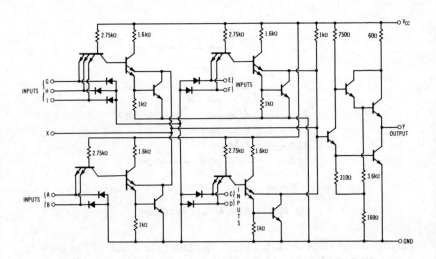

(B) Logic diagram. *(Courtesy Fairchild Camera and Instrument Corp.)*

Fig. 25-3. An expandable 2-2-2-3 AND-OR gate.

Section 26

INVERTERS, BUFFERS, AND DRIVERS

26-1 To Check the Operation of a Hex Inverter Buffer/ Driver

Equipment: Logic pulser and logic probe.

Connections Required: Connect V_{cc} and gnd leads of the pulser and probe to the power supply in the equipment under test (see Fig. 26-1).

Procedure: Inject test pulses at the input of an inverter, and check the output with a logic probe (probe must operate in the TTL mode if the device drives TTL circuitry; the probe must operate in the MOS mode if the device drives MOS circuitry).

Evaluation of Results: In normal operation, an inverter provides a complementary input/output relation. Note that in some digital equipment, to obtain increased fan-out, several inverters in a single package may be paralleled.

26-2 To Check the Operation of a Quad Two-Input NAND Buffer

Equipment: Logic pulser and logic probe.

Connections Required: Connect V_{cc} and gnd leads of the pulser and probe to the power supply in the equipment under test (see Fig. 26-2).

Procedure: Same as in Section 15-3.

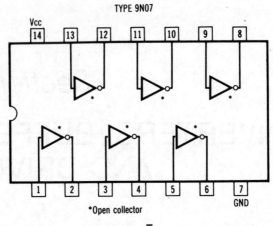

TYPE 9N07

*Open collector

Positive logic: $Y = \overline{A}$

(A) Pinout.

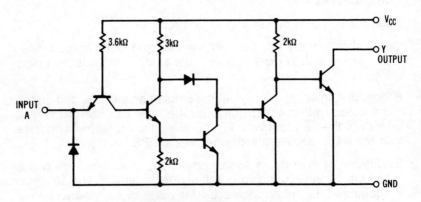

Component values shown are typical.

(B) Logic diagram. *(Courtesy Fairchild Camera and Instrument Corp.)*

Fig. 26-1. A hex inverter buffer/driver.

Evaluation of Results: Same as in Section 15-3. (Current or voltage values may be double or triple the load values used with ordinary NAND gates.)

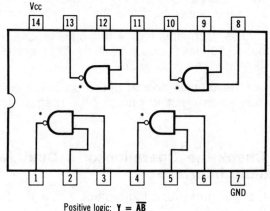

Positive logic: $Y = \overline{AB}$

*OPEN COLLECTOR

(A) Pinout.

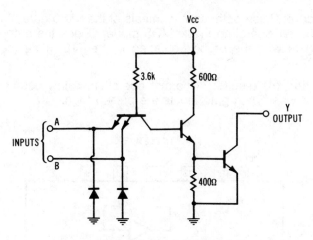

Component values shown are typical.

(B) Logic diagram. *(Courtesy Fairchild Camera and Instrument Corp.)*

Fig. 26-2. A quad 2-input NAND buffer.

26-3 To Check the Operation of a Dual TTL-MOS Interface Element

Equipment: Logic pulser and logic probe.

Connections Required: Connect V_{cc} and gnd leads of the pulser and probe to the power supply in the equipment under test.

Procedure: Pulse each input high with the pulser operating in the MOS mode, and check the output with the logic probe operating in the TTL mode (see Fig. 26-3).

Evaluation of Results: In normal operation a low output is obtained when each input is pulsed high (negative true MOS logic).

26-4 To Check the Operation of a Dual Two-Input Compatible Interface

Equipment: Logic pulser and logic probe.

Connections Required: Connect V_{cc} and gnd leads of the pulser and probe to the power supply in the equipment under test. Tie both input terminals of each NAND gate together. (The expander input need not be included in the test).

Procedure: Pulse both input terminals of the NAND gate high with the pulser operating in the MOS mode. Check the output from the gate with the probe operating in the TTL mode (see Fig. 26-4).

Evaluation of Results: In normal operation, a low output is obtained when both gate inputs are pulsed high.

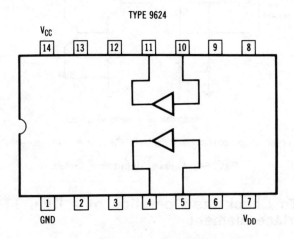

Fig. 26-3. Dual TTL-MOS interface IC.

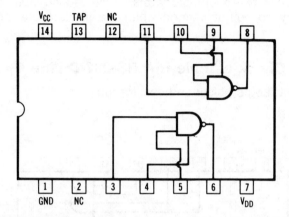

Fig. 26-4. Arrangement of a dual 2-input compatible interface.

26-5 To Check a Dual High Voltage, High Current Driver

Equipment: Logic probe and logic pulser.

Connections Required: Connect V_{cc} and gnd leads of the pulser and probe to the power supply in the equipment under test. Tie the strobe input to a fixed high point. The expander input is automatically verified if the device operates normally. Tie all four inputs together (Fig. 26-5).

Procedure: Inject a logic-high test pulse into the tied inputs. Observe the response at the output terminal with the logic probe.

Evaluation of Results: In normal operation, the output terminal will go low when all four inputs are pulsed high.

26-6 To Check the Operation of a Tri-State Bus Driver

Equipment: Logic pulser and logic probe.

Connections Required: Connect V_{cc} and gnd leads of the pulser and probe to the power supply in the equipment under test (Fig. 26-6).

Procedure: In the first test, tie the enable (control) input to a fixed high-level point; in the second test, tie the enable input to a fixed low-level point. Check output levels with the logic probe.

275

Evaluation of Results: In the first test, the output logic levels normally follow the input logic levels. In the second input, there is normally no output change when the tri-state bus driver is pulsed.

26-7 To Check a Triple EIA RS-232-C Line Driver

Equipment: Logic pulser and oscilloscope.

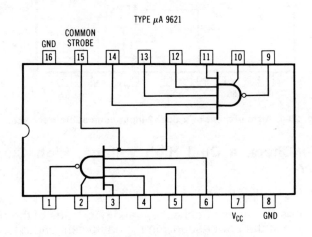

(A) Pinout.

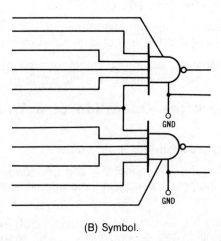

(B) Symbol.

Fig. 26-5 A dual high-voltage

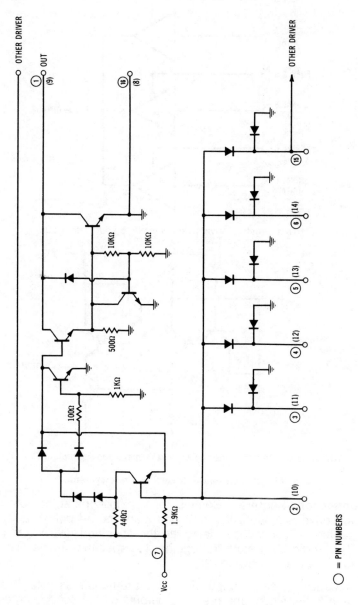

(C) Logic diagram. *(Courtesy Fairchild Camera and Instrument Corp.)*

high-current driver IC.

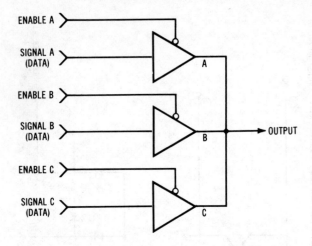

(A) Three tri-state drivers with common outputs into a bus.

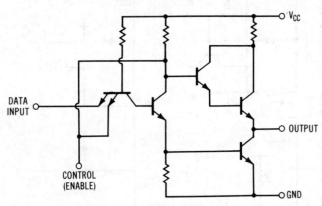

(B) Logic diagram. *(Courtesy Fairchild Camera and Instrument Corp.)*

Fig. 26-6. Tri-state drivers in a bus system.

Connections Required: Connect the oscilloscope at the output of the driver. Connect V_{cc} and gnd leads of the pulser to a 6-V source. In consecutive tests, tie the specified low inputs to a fixed logic-low point. Tie together the specified inputs that are to be pulsed high (Fig. 26-7).

Procedure: Pulse the input and inhibit terminals as listed in the truth table. Check the resulting outputs on the scope screen.

Evaluation of Results: In normal operation, the driver input/output relations will be in accordance with the truth table.

278

26-8 To Check the Operation of a Triple EIA RS-232-C Line Receiver

Equipment: Dual-trace oscilloscope.

Connections Required: Connect the output terminal of the receiver to one input channel of the scope; connect each input terminal of the receiver in turn to the other input channel of the scope (Fig. 26-8).

Procedure: Turn on the equipment for usual operation.

Evaluation of Results: In normal operation, the receiver output is the same as the receiver input, except that the output waveform is inverted.

26-9 To Check the Operation of a Dual Differential Line Driver

Equipment: Logic pulser and logic probe.

Connections Required: Tie the inputs of a driver together in pairs; in the final test, tie all three inputs together. Connect the V_{cc} and gnd leads of the pulser and probe to the power supply in the equipment under test (see Fig. 26-9).

Procedure: First, pulse each individual driver input, checking for output with the probe. Then pulse the driver inputs in pairs, checking for output with the probe. Finally, pulse all three driver inputs simultaneously, checking for output with the probe.

Evaluation of Results: In normal operation, no outputs will be obtained unless all three inputs are pulsed simultaneously.

26-10 To Check the Operation of a True/Complement, Zero/One Element

Equipment: Logic pulser and logic probe.

Connections Required: Connect V_{cc} and gnd leads of the pulser and probe to the power supply in the equipment under test. In the three tests that require a low-level control input (or inputs), tie the associated terminals to the fixed low point (see Fig. 26-10).

Procedure: Pulse the A_1 through A_4 inputs in turn, checking the resulting outputs with the logic probe.

Evaluation of Results: In normal operation, the input/output relations will be in accordance with the listings in the truth table.

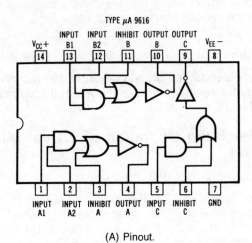

(A) Pinout.

INPUT		INHIBIT	OUTPUT
1	2		
All sections:			
0	0	0	1
1	1	0	0
0	0	1	0
1	1	1	0
For Channels A & B add:			
0	1	0	1
1	0	0	1
0	1	1	0
1	0	1	0

(For Channel C, omit INPUT 2 Column)

(C) Truth table.

Fig. 26-7. An EIA

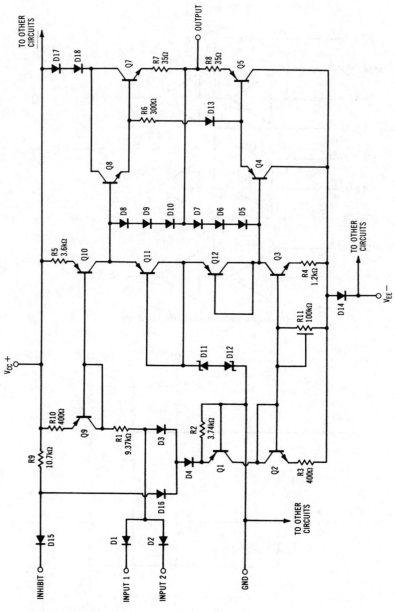

(B) Logic diagram. *(Courtesy Fairchild Camera and Instrument Corp.)*

RS-232-C line driver.

TYPE μA 9617

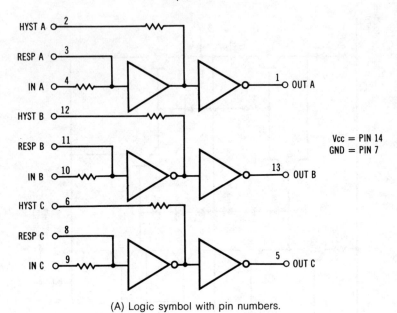

(A) Logic symbol with pin numbers.

Vcc = PIN 14
GND = PIN 7

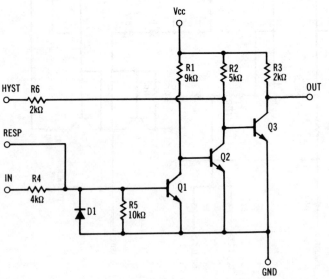

(B) Logic diagram. *(Courtesy Fairchild Camera and Instrument Corp.)*

Fig. 26-8. An EIA RS-232-C line receiver.

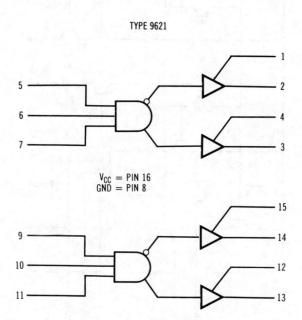

TYPE 9621

V_{CC} = PIN 16
GND = PIN 8

(A) Logic symbol with pin numbers.

Fig. 26-9. A dual differential line driver.

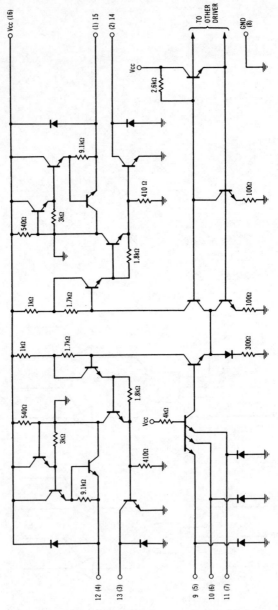

(B) Logic diagram. *(Courtesy Fairchild Camera and Instrument Corp.)*

Fig. 26-9. A dual differential line driver. (Continued)

TYPE 93H87

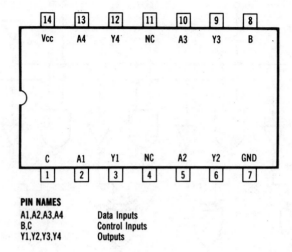

PIN NAMES
A1,A2,A3,A4   Data Inputs
B,C   Control Inputs
Y1,Y2,Y3,Y4   Outputs

(A) Pinout.

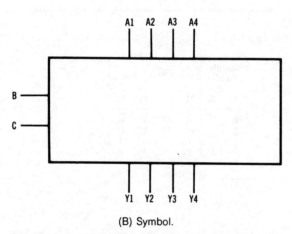

(B) Symbol.

Fig. 26-10. A 4-bit true/complement zero/one element.

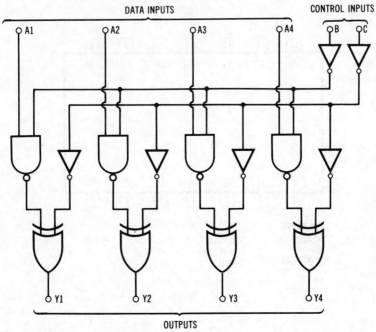

(C) Logic diagram. *(Courtesy Fairchild Camera and Instrument Corp.)*

CONTROL INPUTS		OUTPUTS			
B	C	Y1	Y2	Y3	Y4
0	0	$\overline{A1}$	$\overline{A2}$	$\overline{A3}$	$\overline{A4}$
0	1	A1	A2	A3	A4
1	0	1	1	1	1
1	1	0	0	0	0

(D) Truth table.

Fig. 26-10. A 4-bit true/complement zero/one element. (Continued)

Section 27
MEMORIES

27-1 To Check the Operation of a 256-Bit Read/Write Memory and Decoder Driver

Equipment: Logic comparator.

Connections Required: None.

Procedure: Insert a known good reference IC into the logic comparator and push the test-cable termination over the IC under test. Turn the equipment on as for normal operation.

Evaluation of Results: As the memory is progressively loaded and then unloaded, no discrepancy between the reference IC and the IC under test will normally be indicated by the logic comparator. Make certain that sufficient time is allowed for the memory to become completely loaded, and to become completely unloaded (see Fig. 27-1).

27-2 To Check the Operation of a 64-Bit Fully Decoded Read/Write Memory

Equipment: Logic comparator.

Connections Required: None.

Procedure: Insert a known good reference IC into the logic comparator and push the test-cable termination over the IC under test. Turn the equipment on as for normal operation (see Fig. 27-2).

Evaluation of Results: As the memory is progressively loaded and then unloaded, no discrepancy between the reference IC and the IC under test will normally be indicated by the logic

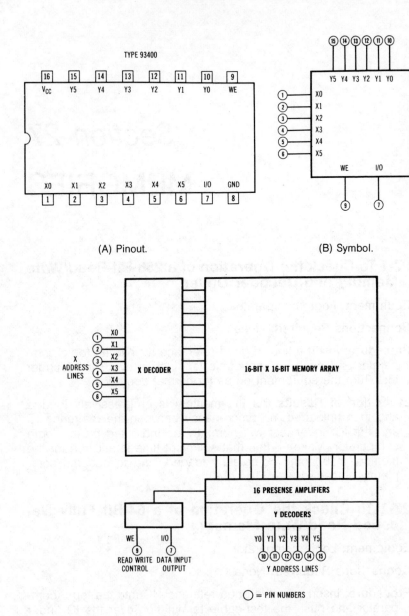

(A) Pinout.

(B) Symbol.

(C) Logic diagram. *(Courtesy Fairchild Camera and Instrument Corp.)*

Fig. 27-1. A 256-bit memory.

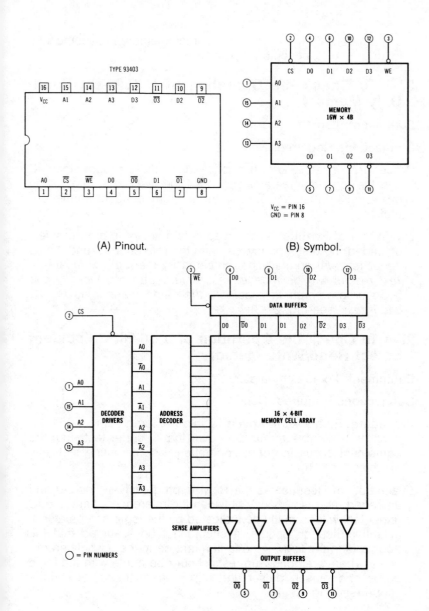

TYPE 93403

(A) Pinout.

(B) Symbol.

V_{CC} = PIN 16
GND = PIN 8

O = PIN NUMBERS

(C) Logic diagram. *(Courtesy Fairchild Camera and Instrument Corp.)*

Fig. 27-2. A 64-bit memory.

comparator. Allow sufficient time for the memory to become fully loaded, and to become fully unloaded.

27-3 To Check the Operation of a 1024-Bit Read-Only Memory

Equipment: Logic comparator.

Connections Required: None.

Procedure: Insert a known good and similarly programmed IC into the logic comparator and push the test-cable termination over the IC under test. Turn on the equipment as for normal operation.

Evaluation of Results: As the read-only memory is progressively unloaded, no discrepancy between the reference IC and the IC under test will normally be indicated by the logic comparator. *Test results will be meaningless unless the reference IC has been programmed in precisely the same manner as the IC under test* (see Fig. 27-3).

27-4 To Check the Operation of a 16-Bit Coincident Select Read/Write Memory

Equipment: Logic comparator.

Connections Required: None.

Procedure: Plug a known good IC into the logic comparator and push its test-cable termination over the IC under test. Turn the equipment on as in normal operation. Use a routine that fully exercises the RAM.

Evaluation of Results: As the RAM is progressively loaded and unloaded, no discrepancy between the reference IC and the IC under test will normally be indicated by the logic comparator. If a fault indication occurs, it should not be assumed that the trouble is in the RAM—the fault might be in the associated circuitry. Therefore, follow-up tests should be made with the logic probe and pulser, pulser and clip, current tracer, or oscilloscope (see Fig. 27-4).

27-5 To Check the Operation of a 1K × 1-Bit Static Ram

Equipment: Logic comparator.

Connections Required: None.

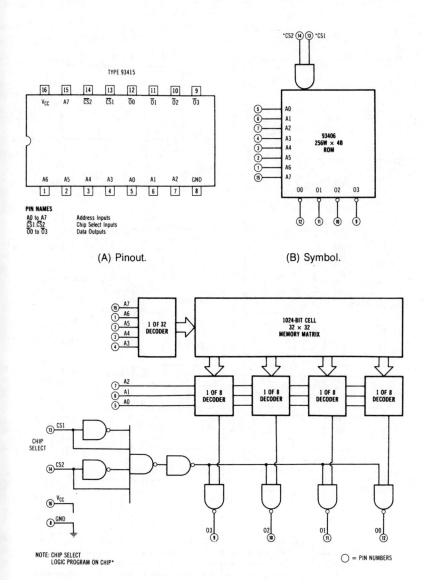

(A) Pinout.

(B) Symbol.

(C) Logic diagram. *(Courtesy Fairchild Camera and Instrument Corp.)*

Fig. 27-3. A 1024-bit ROM.

Procedure: Use a routine that fully exercises the RAM. Plug a known good IC into the logic comparator and push its test-cable termination over the IC under test. Turn the equipment on as in normal operation (see Fig. 27-5).

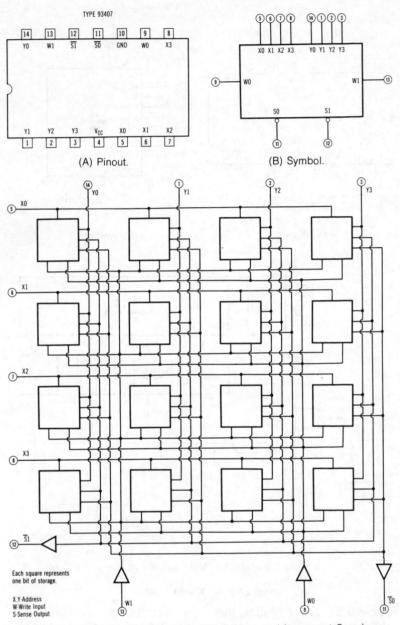

TYPE 93407

(A) Pinout.

(B) Symbol.

(C) Logic diagram. *(Courtesy Fairchild Camera and Instrument Corp.)*

Each square represents one bit of storage.

X-Y-Address
W-Write Input
S-Sense Output

Fig. 27-4. A 16-bit RAM.

Evaluation of Results: No discrepancy between the reference IC and the IC under test will normally be indicated by the logic comparator as the RAM is progressively loaded and unloaded. In case that a fault is indicated by the logic comparator, remember that the trouble may be internal to the RAM, or it may be external in the associated circuitry.

27-6 To Check the Operation of a 4096 × 1-Bit Dynamic Ram

Equipment: Logic comparator.

Connections Required: None.

Procedure: Plug a known good IC into the logic comparator and push its test-cable termination over the IC under test. Use a routine that fully exercises the RAM. Turn the equipment on as in normal operation (see Fig. 27-6).

Evaluation of Results: If the RAM is operating normally, there will be no discrepancy indicated by the logic comparator between the reference IC and the IC under test while the memory is being progressively loaded and unloaded. In case that a fault is indicated by the logic comparator, follow-up tests by the logic pulser and probe, pulser and clip, current tracer, or oscilloscope are generally required.

TYPE 2102A

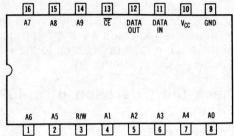

PIN NAMES

D$_{IN}$	Data Input
A0-A9	Address Inputs
R/W	Read/Write Input
$\overline{CE}$	Chip Enable
D$_{OUT}$	Data Output
V$_{CC}$	Power (+5V)

(A) Pinout.

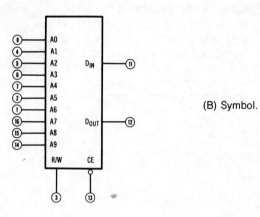

(B) Symbol.

Fig. 27-5. An 1K

294

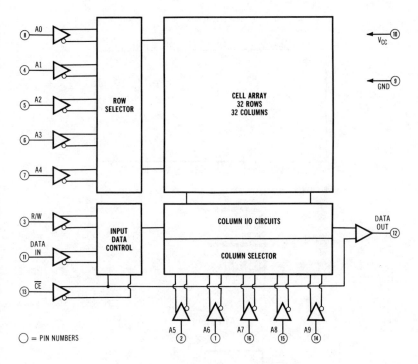

(C) Logic diagram. *(Courtesy Intel Corp.)*

(D) Truth table.

$\overline{CE}$	R/W	D_{IN}	D_{OUT}	MODE
1	X	X	HIGH Z	NOT SELECTED
0	0	0	0	WRITE "0"
0	0	1	1	WRITE "1"
0	1	X	D_{OUT}	READ

X 1-bit RAM.

TYPE 2104A

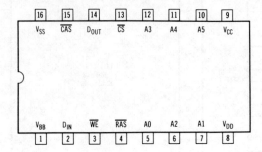

PIN NAMES

A0-A5	Address Inputs
CAS	Column Address Strobe
CS	Chip Select
D$_{IN}$	Data In
D$_{OUT}$	Data Out
RAS	Row Address Strobe
WE	Write Enable
V$_{BB}$	Power (−5V)
V$_{CC}$	Power (+5V)
V$_{DD}$	Power (+12V)
V$_{SS}$	GROUND

(A) Pinout.

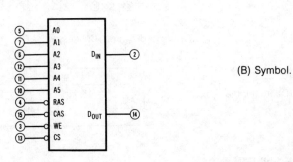

(B) Symbol.

Fig. 27-6. A 4096

296

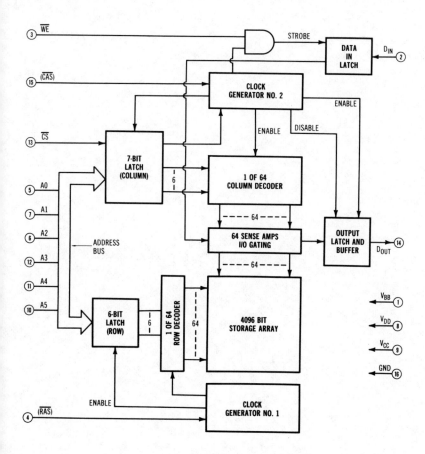

(C) Logic diagram. *(Courtesy Intel Corp.)*

X 1-bit RAM.

Section 28

MISCELLANEOUS DIGITAL TESTS

28-1 To Map the Trouble Area in a Digital System

Equipment: Logic comparator.

Connections Required: None.

Procedure: Note the key signals to be observed in the service manual for the digital equipment. Use the logic comparator to check the ICs in the portion of the circuit that is suspected of failing.

Evaluation of Results: In most trouble situations, the comparator will narrow down the possibilities to a few nodes (sometimes to a single node). An experienced digital technician can check out an IC in less than half a minute. In some situations, various ICs may not be testable with a logic comparator. Therefore, input/output relations must be checked out with a logic probe. A logic pulser may be used to replace the clock and to single-step the circuit action.

28-2 To Make a Temperature Test of an IC

Equipment: Electronic thermometer.

Connections Required: None.

Procedure: Measure the surface temperature of the suspected IC and that of a similar IC.

Evaluation of Results: When an IC is shorted internally, more power is dissipated than in normal operation. In turn, the sur-

299

face temperature of a shorted IC is comparatively high. An external short, as between pc conductors, can also cause an IC to run hot because of excessive current demand.

28-3 To Identify a Defective IC by Voltage Measurements

Equipment: Digital voltmeter.

Connections Required: None.

Procedure: Measure the V_{cc} supply voltage, then measure the voltages in the circuit(s) suspected of malfunctioning.

Evaluation of Results: High-level voltage values in digital networks are never quite as high as V_{cc} (in normal operation). For example, in a TTL circuit with V_{cc} = 5 V, high levels through the circuit normally range from 3.5 to 4.5 V. If a heavy load is being driven, the high level might be down to 2.4 V. If a high level is measured equal to V_{cc}, look for a short circuit between the output pin and V_{cc}. Low-level voltage values in digital networks are never zero (in normal operation). Unless a few hundred millivolts is measured at a low output, look for a short circuit between the output pin and ground.

28-4 To Localize a Short Circuit With a Milliohmmeter

Equipment: Digital voltmeter with milliohmmeter function.

Connections Required: None.

Procedure: Measure the resistance between pc conductors in the trouble area. Then move the test prods farther down the circuit and measure the resistance again; if a higher value is measured, move the test prods along the conductors in the opposite direction and measure the resistance again.

Evaluation of Results: If the milliohmmeter reading increases along a defective circuit, the test points are being taken farther from the short-circuit point. On the other hand, if the milliohmmeter reading decreases along a defective circuit, the test points are being taken toward the short-circuit point. The short circuit will be located close to the point of minimum resistance.

28-5 Checking Circuit Action by Pin Lifting

Equipment: Logic probe and logic pulser.

Connections Required: Connect V_{cc} and gnd leads of the pulser and probe to the power supply in the equipment under test.

Procedure: Pin lifting consists of disconnecting an IC from its socket during test procedures. That is, an IC is unplugged from its socket, and a particular pin is bent aside; the IC is then replaced in its socket. Long-nose pliers are generally used to bend IC pins; a 45° bend is adequate. Check the circuit action with the pulser and probe, taking into account the lifted pin(s).

Evaluation of Results: Test results that formerly did not "make sense" may become clarified when IC pins are lifted. For example, glitches are sometimes fed back to a preset or clear pin by a defective IC in another part of the circuit; glitches can greatly confuse a trouble analysis. It may be permissible to "float" a lifted IC pin. On the other hand, a lifted pin may have to be tied to a fixed high or low source in order to obtain a valid test of circuit action.

28-6 Trouble Analysis by Card Swapping

Equipment: None.

Connections Required: None.

Procedure: It is sometimes possible to get a "handle" on a trouble symptom by card swapping; that is, two pc boards in a modular system are interchanged, and the circuit action that results is observed.

Evaluation of Results: The troubleshooter may evaluate card swapping tests in either one of two ways: First, he or she may be guided by case histories; second, he or she may have an adequate understanding of system operation to correctly interpret the results of card swapping. Indiscriminate card swapping can be a disastrous procedure; before any pc boards are interchanged, the technician must be certain that he or she fully understands the changes in circuit configurations and the resulting signal responses that will occur. If card-swapping is done knowledgeably, useful preliminary troubleshooting data can frequently be obtained.

28-7 Troubleshooting by Comparative Tests

Equipment: Logic probe, logic pulser, logic clip, current tracer, logic comparator, and a similar unit of digital equipment in normal operating condition.

Connections Required: None.

Procedure: The digital equipment under test and the reference unit are first compared with respect to their resting states. Then, a simple routine is run on each unit, and the logic states at various test points are compared. Additional routines may be checked out on both units in a similar manner.

Evaluation of Results: Comparative tests are useful in "tough-dog" situations because no service manual can be completely comprehensive. Comparative tests are also helpful when the technician does not have a complete understanding of system operation and must adopt a "shotgun" approach.

28-8 To Check the Output From a Logic Pulser

Equipment: Oscilloscope with triggered sweep.

Connections Required: Connect V_{cc} and gnd leads of the pulser to a battery with voltage equal to the rated V_{cc} value of the pulser.

Procedure: Apply the output of the pulser to the vertical-input channel of the scope. Activate the pulser.

Evaluation of Results: The peak-to-peak voltage of the pulser signals and the signal waveforms can be checked on the scope screen. This test provides a useful checkup if it is suspected that the pulser is operating marginally.

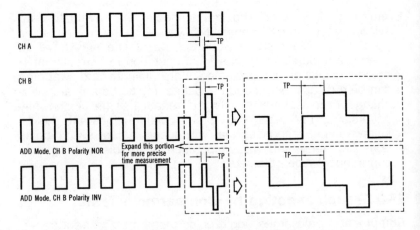

Fig. 28-1. Measurement of propagation delay time.
(Courtesy B&K Precision Dynascan Corp.)

28-9 To Measure Propagation Delay

Equipment: Triggered-sweep oscilloscope, preferably with A+B and A−B display modes.

Connections Required: Connect the input of the configuration (such as a divide-by-eight arrangement) to channel A of the scope, and connect the output of the configuration to channel B.

Procedure: Obtain a convenient A and B display. Then switch the scope operation to A+B or to A−B mode.

Evaluation of Results: With reference to Fig. 28-1, the propagation delay T_p is displayed as a pulse in the A−B display. When expanded, the elapsed time (propagation delay) represented by the pulse can be precisely measured.

28-10 To Check the Output From a Digital Word Generator

Equipment: Word generator.

Connections Required: Connect the serial output to vertical-input channel A; connect the clock output to vertical-input channel B of the scope.

Procedure: Set the generator for repetitive output of a chosen digital word.

Evaluation of Results: The peak-to-peak voltage, waveform, and binary meaning of the generator signal can be observed on the scope screen. If parallel output from the generator is to be checked, the output waveforms must be displayed in pairs.

28-11 To Check Master and Subordinate Clock Signals

Equipment: Dual-trace triggered-sweep oscilloscope.

Connections Required: Connect the master clock line to vertical-input channel A; connect the subordinate clock line to vertical-input channel B of the scope.

Procedure: Observe the timing relation between the two patterns.

Evaluation of Results: With reference to Fig. 28-2, one standard master-subordinate clock relation consists of a 180° phase difference between the two waveforms. Note that if a short-circuit fault occurred in the inverter, the two waveforms would

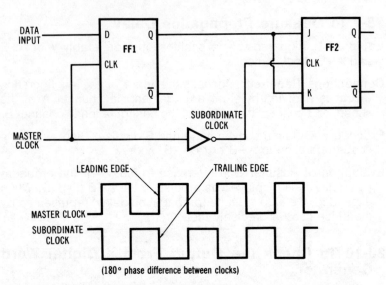

Fig. 28-2. A flip-flop circuit using master and subordinate clock signals.
(Courtesy Hewlett-Packard Co.)

have zero phase difference, and the digital system would present a tough-dog trouble symptom.

INDEX

306